Microbial Extrachromosomal Genetics

Aspects of Microbiology

Series editors
J. A. Cole, University of Birmingham
C. J. Knowles, University of Kent
D. Schlessinger, Washington University School of Medicine, USA

The American Society for Microbiology is now publishing a series of short books. We want to specify the rationale and nature of the series.

When students enroll in college science courses, they are customarily faced with large textbooks that build on secondary sources. Much of the text repeats simple material that has already been covered in high school courses, and the work in active fields is so quickly dated that the treatment in an 'inclusive' book is at best uneven. It is most discouraging that the evidence for critical inferences, the definition of the present limits of knowledge, and the excitement of scientific research are usually denied to all but the few students who go on to graduate work. Undergraduates often can only wonder what their professors are so excited about.

The disparity between science at the 'frontier' and the compilations in textbooks has led some to use collections of 'seminal papers' as a teaching aid. However, these lack continuity and clear expository prose.

The alternative that we are sponsoring is to select a number of the liveliest topics and ask active researchers who also write well to provide short books like this one. In addition to references to recent studies, each book provides a précis of the state of the field, providing the background necessary to bring students to the heart of the science. We expect these books to supplement a course, to provide additional material for undergraduate and graduate students, or to provide the complete or partial basis for all courses on microbiology, molecular biology, microbial ecology, applied microbiology, medical microbiology, etc.

With experience in providing books and scientific literature to an international audience of students and scientists, the ASM has agreed to copublish the United States editions with Van Nostrand Reinhold (UK).

David Schlessinger
ASM Series Editor

1. Oral microbiology (2nd edition) *P. D. Marsh and M. V. Martin*
2. Bacterial toxins *J. Stephen and R. A. Pietrowski*
3. The microbial cell cycle *C. Edwards*
4. Bacterial plasmids *K. Hardy*
5. Bacterial respiration and photosynthesis *C. W. Jones*
6. Bacterial cell structure *H. J. Rogers*
7. Microbial control of plant pests and diseases *J. W. Deacon*
8. Methylotrophy and methanogenesis *P. J. Large*
9. Extracellular enzymes *F. G. Priest*
10. Intestinal microbiology *B. S. Drasar and P. A. Barrow*
11. Biotechnology principles *J. E. Smith*
12. Microbial extrachromosomal genetics *S. G. Oliver and T. A. Brown*

Aspects of Microbiology 12

Microbial Extrachromosomal Genetics

Stephen Oliver
Senior Lecturer in Applied Molecular Biology
University of Manchester Institute of Science and Technology

Terence Brown
Lecturer in Biotechnology
University of Manchester Institute of Science and Technology

American Society for Microbiology

Published in 1985 by Van Nostrand Reinhold (UK) Co. Ltd
Molly Millars Lane, Wokingham, Berkshire, England

Published in the USA by
American Society for Microbiology
1913 I Street, N.W., Washington, D.C. 20006, USA

Photoset in Times 9 on 10pt by Kelly Typesetting Limited
Bradford-on-Avon, Wiltshire

Printed and bound in Hong Kong

ISBN 0-914826-79-4

Preface

The eukaryotic cell is a genetic chimera; it contains structurally and functionally distinct genomes in its nucleus, cytoplasm and cytoplasmic organelles. The phenotype of a eukaryote is dependent on contributions from all of these genomes and on cooperation between them. The elucidation of these separate contributions and their interactions is a fascinating problem in cell and molecular biology, and one in which microorganisms represent important experimental subjects. This book traces the development of our understanding of extra-chromosomal genetics in microorganisms from its beginning in the discovery of anomalous, and even bizarre, inheritance patterns to the mechanistic molecular biological analyses of today. It gives special emphasis to mitochondrial and chloroplast inheritance, discusses the evolutionary origin of these organelle genomes and relates this to the development of the eukaryotic state. We hope that this volume will be of use to students of microbiology, genetics, biochemistry, molecular biology and evolutionary biology. In such a brief work we have necessarily made omissions and simplifications, but we hope that some readers will be stimulated to undertake a deeper study of this topic and we have offered some guidance on further reading.

This work has benefited from the criticisms of the series editors and of the referee nominated by the American Society for Microbiology. Most of all it has profited from the continual editorial comment of Rowena Oliver whose insistence that it should be understandable to a zoologist has, we trust, decreased its opacity.

S. G. OLIVER, T. A. BROWN
Manchester

Contents

1 Introduction

Gregor Mendel was either very shrewd or very fortunate in his choice of characters when he studied inheritance in the garden pea. The characters that he chose, such as seed colour and seed shape, proved to be controlled by genes which were carried on different nuclear chromosomes. The behaviour of these genes during mating experiments therefore reflected the behaviour of chromosomes during meiosis (the formation of germ cells) and zygosis (the fusion of the haploid nuclei of germ cells). Thus, Mendel's laws of segregation and independent assortment were a formal description of chromosome behaviour. There are, however, genes which do not obey Mendel's laws because they are not borne on nuclear chromosomes. These are known as extrachromosomal or non-Mendelian genes.

The first description of extrachromosomal inheritance was made by one of the rediscoverers of Mendel's work, the German botanist, Carl Correns. In his pioneering study, Correns employed the flowering plant *Mirabilis jalapa* in his breeding experiments. This is a variegated plant which originated in South America (hence its common name, the Marvel of Peru) and is grown as a perennial in European gardens. The leaves of *Mirabilis* may be of three colour types, green, pale or variegated, and Correns' experiments were designed to study the inheritance of this character. His results, which were published in 1909, are summarized in Table 1. This shows that the inheritance of leaf colour in this plant breaks one of the rules of Mendelian genetics, that the results of reciprocal crosses should be identical. Irrespective of whether a given allele (for instance, that for pale leaves) is donated by the male or the female partner in a cross, its pattern of inheritance should always be the same. Correns' results showed that in *Mirabilis* the contribution of the male partner, the pollen grains, to the inheritance of leaf colour was nil. This trait was solely determined by the female partner. This phenomenon is known as maternal inheritance and is a common one in extrachromosomal genetics. It should be noted, however, that not all cases where different patterns of inheritance are found in reciprocal crosses are due to

Table 1 Inheritance of leaf colour in *Mirabilis jalapa*

Pollen from branch with leaves of type	Pollinate flowers from branch with leaves of type	Progeny have leaves of type
	Pale	Pale
Pale	Green	Green
	Variegated	Pale, green and variegated
	Pale	Pale
Green	Green	Green
	Variegated	Pale, green and variegated
	Pale	Pale
Variegated	Green	Green
	Variegated	Pale, green and variegated

extrachromosomal genes. For instance, sex-linked genes in animals show such non-reciprocity. Therefore the inheritance pattern of a given character must satisfy a number of criteria before we can be confident that it is determined by an extrachromosomal gene. These will be dealt with in more detail later in the chapter.

Since the pollen grain in *Mirabilis* donates little or no cytoplasm to the zygote cell and since it also has no influence on the inheritance of leaf colour it may be concluded that this trait is controlled by a cytoplasmic determinant. The colour of the leaves is determined by the colour of the chloroplasts which they contain. It is now known that the chloroplast is a cytoplasmic organelle which contains a distinct form of DNA (deoxyribonucleic acid) which differs in both its genetical and physical characteristics from the nuclear DNA of the same cell. Moreover, the chloroplast contains its own machinery for the expression, via the processes of transcription and translation, of the genes carried by that DNA.

Chloroplasts are not unique in having their own genetic system distinct from that of the nucleus. This is also true of the mitochondria, a class of cytoplasmic organelle found in all eukaryotic cells. Although chloroplast and mitochondrial DNA encode certain proteins which are essential for the normal function of their respective organelles, they do not contain enough information for the complete fabrication of chloroplasts and mitochondria (Table 2). The construction of these two kinds of organelle is therefore a cooperative effort by the nuclear and the

Table 2 General features of the mitochondria and chloroplasts of eukaryotic microbes, with particular reference to their genetic systems

	Mitochondria	Chloroplasts
Major biochemical functions	Oxidation of carbohydrates, lipids and amino acids, involving the enzymes of the electron transport chain, and resulting in ATP synthesis by oxidative phosphorylation.	Photosynthesis, fatty acid synthesis, nitrite reduction.
Approximate range of genome sizes	20–150 kb	85—possibly 2000 kb
Approximate number of genes carried by the genome.	rRNA: 2 tRNA: 25–35 protein: 10–20	3–5 about 35 50–100+
Proteins known to be encoded by the organellar genome	Mostly components of the electron transport chain: apocytochrome b cytochrome c oxidase (3 subunits) ATPase (2–3 subunits) also 1 ribosomal protein	Many proteins directly involved in photosynthesis: 12–15 thylakoid membrane proteins ribulose biphosphate carboxylase (large subunit) 5 ATPase subunits Also 20–21 ribosomal proteins Protein synthesis factors (e.g. EF–Tu).

organellar genomes. There are also other sites of genetic information within the cytoplasm of eukaryotic cells and the phenotype of such a cell is a composite expression of its nuclear and cytoplasmic genomes. The interaction of these two during growth and division is a fascinating problem in cell biology.

Extrachromosomal genetics has been studied in all kinds of eukaryotes, but microorganisms have become favoured experimental subjects in this branch of genetics as in so many others. They have many features which facilitate genetic studies. Large populations may easily be handled and this permits the induction and selection of rare mutations. Haploid, uninucleate forms occur in some stage of the life cycle of almost all microorganisms which makes detection of nuclear mutations easier. Many eukaryotic microorganisms can be grown and manipulated in the same manner as bacteria, they can be grown in defined media, on agar plates and their colonies may be replicated using velvet pads. All of these features facilitate genetic analysis.

The structural simplicity of many microorganisms commends their use in the study of extrachromosomal genetics. For instance, the unicellular alga, *Chlamydomonas reinhardtii*, has only a single chloroplast and this is one reason for its popularity in the study of chloroplast genetics. The metabolic versatility of microorganisms again commends them for this kind of work. The ability of many green algae to grow heterotrophically on simple carbon substrates as well as autotrophically by photosynthesis is very useful when isolating mutants defective in chloroplast functions. Correspondingly, the budding yeast, *Saccharomyces cerevisiae*, can grow either fermentatively or oxidatively making it the organism of choice for many studies of mitochondrial inheritance. The novel life cycles of a number of microorganisms are also useful. For instance, the heterokaryon stage of many filamentous fungi, in which the cytoplasms of each of the two parents in a cross are mixed but there is no fusion of parental nuclei, may be used to determine whether a given gene is truly cytoplasmic or merely extrachromosomal.

No system is perfect, and perhaps the biggest drawback to the use of microorganisms for these kinds of studies is that many of them have tough cell walls requiring quite harsh methods to break open the cells. This makes the isolation of intact cytoplasmic organelles difficult and cytoplasmic genetics is mainly concerned with organelle genetics.

This brief introduction has given some idea about the nature of extrachromosomal genetics and the reasons why microorganisms are so useful in its study. The rest of the chapter will be devoted to a detailed consideration of the criteria which permit the assignment of a given gene to an extrachromosomal location and to the features of various microorganisms which are of use in making such an assignment.

The first point to make is that we cannot write a list of rigid criteria, all of which must be satisfied in order to assign a particular gene to an extrachromosomal location. Rather, we can supply a list of diagnostic features, more than one of which must generally apply in order to make a confident assignment.

Differences in inheritance patterns between reciprocal crosses

Non-reciprocity has already been discussed with reference to chloroplast inheritance in *Mirabilis*. It was noted then that sex-linked genes in higher organisms also give differing patterns of inheritance in reciprocal crosses and this emphasizes the fact that this list of features is diagnostic rather than definitive. Correns' experiments with *Mirabilis* were the first published demonstration of maternal inheritance and this is the most common non-reciprocal effect associated with non-Mendelian inheritance.

Maternal inheritance can still occur when the different mating types of a species are not physically distinguishable. Such isogamous matings are quite common among microorganisms and an important example of this phenomenon is chloroplast inheritance in *Chlamydomonas reinhardtii*. In this organism mating is isogamous and there are two mating types, *mt*+ and *mt*−. In crosses, alleles of chloroplast genes carried by the *mt*+ parent are transmitted to all of the progeny of the cross, the *mt*− parent making no contribution. This is an example of maternal inheritance in which the sexes are not morphologically distinct and where there is apparently complete cytoplasmic mixing in the zygote.

Non-Mendelian segregation at meiosis

The rules of Mendelian genetics, as noted earlier, may be regarded as a formal description of chromosome behaviour at meiosis. Many fungi may be used to examine directly the results of meiotic segregation since in these species meiosis produces haploid spores from which vegetative colonies may be derived. The set of four (or multiples of four) spores produced in this manner are referred to as a meiotic tetrad. Two alternate alleles of a chromosomal or Mendelian gene will segregate 2 wild type:2 mutants in such a tetrad assuming that any recombination that occurs is reciprocal. Non-Mendelian genes will not segregate in this manner and therefore a 4 wild type:0 mutants or 0 wild type:4 mutants ratio in the meiotic tetrad is strongly indicative of an extrachromosomal location for the gene being studied.

Somatic or mitotic segregation

The whole purpose of the complex process of mitotic nuclear division is to produce two daughter nuclei which are genetically identical. Therefore there should be no segregation of the alleles of a chromosomal gene during vegetative growth. When different allelic forms do segregate out during mitosis, for instance to produce sectored colonies, then it is likely that the gene concerned is not carried on a nuclear chromosome but is cytoplasmically located. Again, caution should be exercised because sectored colonies can arise due to high frequencies of reversion to wild type or from non-reciprocal recombination events in diploid nuclei.

Gene transfer in the absence of nuclear fusion

This is a particularly useful test since it can show whether a given determinant is not only extrachromosomal but is also extranuclear. The demonstration of gene transfer without the fusion of nuclei is most readily achieved in the fungi. This can be done by following the 'invasive' or 'infective' spread of a character during mating when cell fusion but not nuclear fusion has occurred. In such an experiment a character carried by only one parent can be retrieved by culturing parts of the mycelium of the other parent which were not directly involved in hyphal fusion and nuclear exchange. An example is the infective spreading of senescence between mycelia of the ascomycete, *Podospora anserina*.

The second method of demonstrating gene transfer in the absence of nuclear fusion that may be exploited in fungi is the establishment of a heterokaryon. In many members of the higher fungi, the *Ascomycetes* and *Basidiomycetes*, the mating of two compatible haploid mycelia leads to plasmogamy, total cytoplasmic mixing and nuclear migration. Thus, in the new mycelium formed, which is called a heterokaryon, the two parental types of nuclei coexist in a common cytoplasm but never, or very rarely, fuse and thus cannot exchange genes (Figure 1). In the *Ascomycetes* the proportions of the two kinds of nuclei in different parts of the heterokaryotic mycelium may vary. However, in the *Basidiomycetes* a dikaryon is formed where the hyphae are divided by special septa into segments which each contain one of each parental type of nucleus. Heterokaryons produce vegetative spores which contain a single nucleus of either parental type surrounded by the common cytoplasm. Therefore, any character donated by only one of the parents to the heterokaryon, which appears in all the progeny grown from its spores must have been transmitted through the cytoplasm.

Absence of linkage to known nuclear genes

The nuclear chromosomes of most eukaryotes can be visualized cytologically and also defined genetically as a number of linkage groups. However, the chromosomes of a number of microfungi such as *Saccharomyces* (yeast) and *Ustilago* (smut) have never convincingly been demonstrated microscopically but it is always possible to assign any nuclear gene to the linkage group which defines, in genetic terms, the chromosome on which it lies. It follows that an extrachromosomal gene should not show any linkage to known nuclear genes. Such lack of linkage is readily demonstrated in species which have a well-defined nuclear genetic map in which markers in the centromere-linked genes of all the chromosomes are available. This emphasizes the fact that it is much easier to do extrachromosomal genetics in species in which many of the chromosomal genes have been characterized.

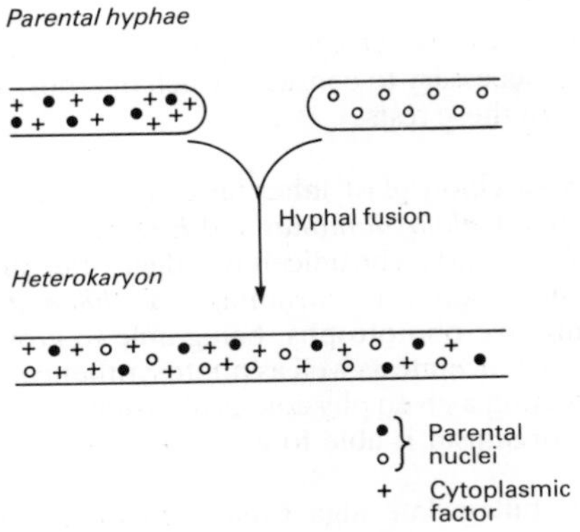

Fig. 1 Heterokaryon formation.

Specific mutational effects

A suggestive, but by no means conclusive, indication that a given mutation is a change in a non-Mendelian genome is given if that mutation can be induced by an agent which is not known to affect nuclear genes. One example is the induction of mutations in the chloroplast genome of *Chlamydomonas* by streptomycin, an inhibitor of protein synthesis on 70S ribosomes, or fluorodeoxyuridine. Others are the induction of the cytoplasmic petite mutation in yeast by the DNA intercalator ethidium bromide or the uracil analogue 5-fluorouracil and the induction of more limited mutations in the yeast mitochondrial genome by Mn^{2+} ions. The specificity of such an effect is confirmed if the agent induces mutations leading to a loss of function in strains which are nuclear diploids or tetraploids. A null mutation in a nuclear gene would be recessive and so not phenotypically expressed in such strains.

Correlation of a given character with the presence of some physical entity distinct from the nuclear chromosomes

In most cases an extrachromosomal genome is first defined genetically and then a search is made for its physical basis. The killer phenomenon in *Paramecium* was recognized as a heritable characteristic and later correlated with the presence of the kappa particles in the cytoplasm. Similarly, the cytoplasmic petite mutation in yeast was well known before being correlated with the loss or severe alteration of mitochondrial DNA. Occasionally biochemists discover a self-replicative molecule in the cytoplasm before geneticists have recognized any phenotype with which it is associated.

This is the case with the 2 μm circular DNA of yeast. On other occasions it is the biochemists who fail the geneticists. For instance, the *psi* factor of yeast, a non-Mendelian determinant which modifies the activity of suppressor genes, has yet to be assigned to any self-replicative molecule.

Experimental organisms

Now that the various tests which can be applied to extrachromosomal genes have been discussed it is necessary to consider which microorganisms are most suitable for the application of these tests.

Algae The study of chloroplast inheritance has been largely confined to unicellular algae, such as *Chlamydomonas* and *Euglena*, and to variegated flowering plants, such as *Pelargonium*. The unicellular algae offer all the usual advantages of microorganisms and a significant advantage of *Chlamydomonas* and *Euglena* is that they are facultative phototrophs, being able to grow in the dark on simple carbon sources such as acetate as well as photosynthetically in the light. To obtain null mutations affecting a given physiological system (in this case photosynthesis) it is helpful if the organism is able to grow and divide even when the system is inactive.

The unicellular, biflagellate alga *Chlamydomonas reinhardtii* has been the principal microorganism used for the study of chloroplast genetics and its life cycle will be considered in some detail. When grown in liquid culture *Chlamydomonas* is a motile cell bearing two, equal-sized flagella. However, when grown on an agar plate the cells lose their flagella and divide to form discrete colonies. The cell has a rigid cellulose wall which must be ruptured before subcellular organelles or high molecular weight nucleic acids can be isolated. It has a large single chloroplast which cups the nucleus of the cell. Genetic analysis has defined 17 linkage groups within this nucleus, but microscopic visualization of chromosomes is difficult, as it is with many microbial eukaryotes. The nuclear genetics of *Chlamydomonas* is quite well developed. It was the first organism to be used in tetrad analysis (by Pascher in 1916), but the study of chloroplast genetics, initiated by Sager in the early 1950s, has been the real impetus behind the development of its nuclear genetic system. More recently, *Chlamydomonas* has been used to study the genetic control of the cell cycle and this has led to an even more detailed analysis of its nuclear genome.

The life cycle of *Chlamydomonas reinhardtii* is shown in Figure 2. If the motile, vegetative cells are deprived of nitrogen for 2–4 hours, they differentiate into gametes and when gametes of opposite mating types (mt^+ and mt^-) are mixed, they immediately clump together. Pairs of cells of opposite mating type then fuse to produce binucleate, tetraflagellate zygotes. When this mixture is then plated onto agar no further matings are initiated but the zygotes which have already formed lose their flagella and lay down a thick zygospore wall, at the same time doubling their size. Nuclear fusion follows 3.5 hours later after which meiosis occurs, taking about 12 hours. Spore maturation is completed in approximately 6 days and the mature spores will germinate if transferred to fresh agar medium and placed in the light. The zygote cell wall ruptures to release either 4 or 8 haploid cells depending on the strain used. The meiotic tetrad may be analysed either by a 'self-dissection' process which uses water drops to disperse the four spores or by more conventional micromanipulation.

Nuclear genes, such as the mating type locus, segregate 2:2 in this meiotic tetrad. Chloroplast genes are inherited in a uniparental fashion reminiscent of

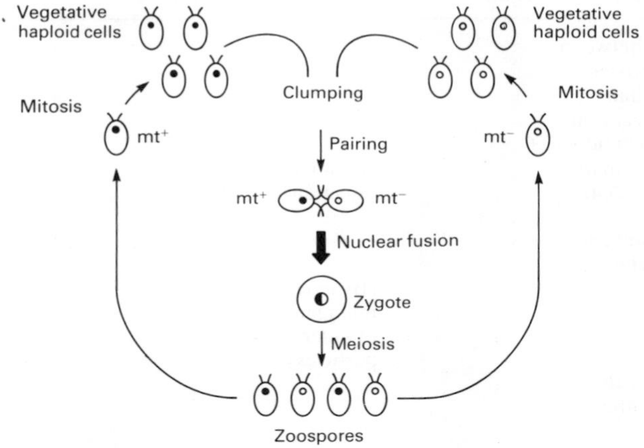

Fig. 2 Life cycle of *Chlamydomonas reinhardtii.*

maternal inheritance in higher plants. In over 90% of tetrads the chloroplast genotype of the mt^+ parent, hence referred to as the maternal partner, is inherited by all the four spores in the tetrad. This feature greatly facilitates the identification of the chloroplast genes. Unfortunately, it also prevents their further genetic analysis because no recombinants may be formed between the maternal (mt^+) and paternal (mt^-) genomes. Some means were needed for increasing the occurrence of the rare phenomenon of biparental transmission in which the zygotes produced from a cross can give rise, by meiosis, to haploid cells containing both the maternal and paternal chloroplast genomes. The two different genomes may recombine and segregate during the succeeding mitotic divisions. Cells containing both parental chloroplast genomes are hence cytoplasmic heterokaryons, and are also known as heteroplasmons or cytohets. Sager and Raminis found that the proportion of biparental zygotes could be increased from much less than 10% to about 50% by the irradiation of the mt^+ parents with ultraviolet light immediately before mating. Heteroplasmons are therefore produced at sufficient frequency to permit the study of chloroplast recombination and the construction of genetic maps for the plastid genome. Sager also found that mutations could be selectively introduced into the chloroplast genome by using high concentrations of either fluorodeoxyuridine or streptomycin so that a detailed genetic map could be constructed.

Table 3 shows that six of the seven diagnostic tests for the identification and analysis of extrachromosomal genes may be applied to the chloroplast system of *Chlamydomonas* which make it a near-perfect system for studying extrachromosomal genetics; considerable progress has been made with this organism.

Protozoa The protozoa are difficult organisms with which to work, mainly because bacteriological techniques are not easily applied to them. Nevertheless,

Table 3

Diagnostic test	*Chlamydomonas reinhardtii*	*Paramecium aurelia*	*Saccharomyces cerevisiae*	*Neurospora crassa*
Differences between reciprocal crosses	+	−	−	+
Non-Mendelian meiotic segregation	+	−	+	+
Mitotic segregation	+	+	+	+
Gene transfer in the absence of nuclear fusion	−	(By micro-injection)	±	+
Absence of linkage to nuclear genes	+	−	+	+
Specific mutational effects	+	(By *in vitro* mutagenesis of killer particles)	+	+
Correlation with some entity other than nuclear chromosomes	+	+	+	+

because they are 'true' animals, they may be claimed with some justification to represent a better model for the human cell than most microorganisms. Moreover, since most protozoa do not possess a rigid cell wall, subcellular organelles may be isolated from them with relative ease.

The ciliate, *Paramecium aurelia*, has been very important in the study of extrachromosomal genetics. This organism, like many protozoans, contains both germ-line and somatic nuclei. It has a single macronucleus, containing several hundred times the normal haploid complement of DNA and this controls general growth and development. Each *Paramecium* also contains two diploid micronuclei. These are the germ-line nuclei, involved in sexual reproduction, and giving rise to the somatic macronucleus.

Sexual reproduction in *Paramecium aurelia* occurs by conjugation between individuals of compatible mating type: the process is outlined in Figure 3a. Sexually compatible strains of *Paramecium* define a number of subspecies known as syngens. *Paramecia* may be induced to conjugate by nutrient deprivation. Contact between a pair of compatible cells (gamonts) initiates the breakdown of their respective macronuclei while each of the two micronuclei in each cell undergo meiosis to produce a total of eight haploid nuclei per gamont. Seven of these nuclei then disintegrate. The remaining nucleus in each cell undergoes a single mitotic division and one daughter nucleus from each gamont migrates into

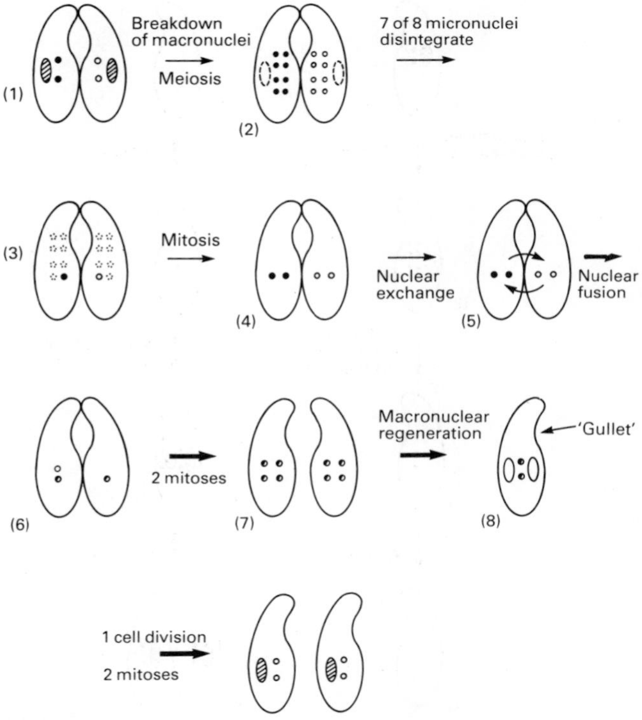

Fig. 3a Conjugation in *Paramecium aurelia.*

the cytoplasm of the other partner through cytoplasmic bridges which have been established in the gullet region (Figure 3a). Each of the migratory nuclei then fuses with the stationary nucleus in its partner gamont to re-establish the diploid number of chromosomes. In addition to this exchange of nuclei during conjugation there may, in some strains, be a fairly extensive exchange of cytoplasm between the mating partners if they are not immediately separated after nuclear fusion.

Separation of the mating partners is followed by two rounds of mitotic nuclear division giving four diploid nuclei, two of which develop into macronuclei and two remain as micronuclei. The two micronuclei each undergo a further mitotic division which is followed by a cell division. The two daughter *Paramecia* now have the usual complement of two diploid micronuclei and one macronucleus.

Paramecium may by-pass the sexual process of conjugation by a parasexual mechanism known as autogamy (see Figure 3b). Autogamy is essentially the same as conjugation except that two haploid nuclei are generated and fuse within a single cell, no sexual partner being involved. For the geneticist, it is a useful method of rapidly establishing homozygous cell lines.

Reference to Table 2 shows that *Paramecium* is not such an amenable subject

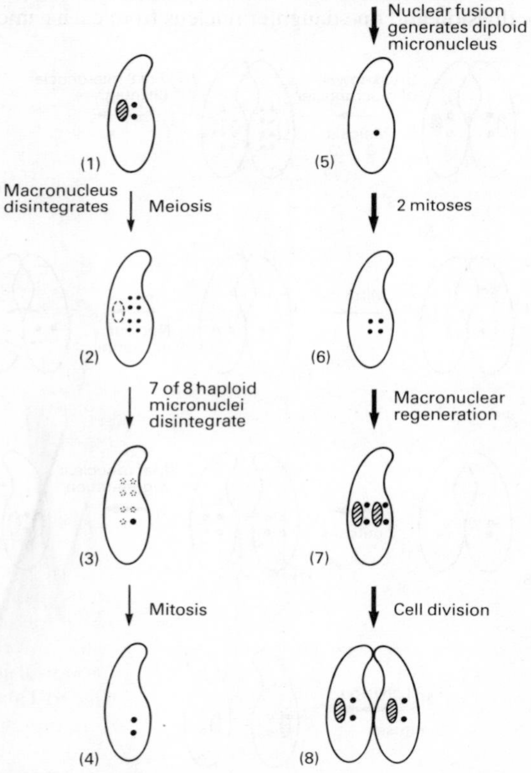

Fig. 3b Autogamy in *Paramecium aurelia.*

for research into extrachromosomal genetics as *Chlamydomonas*. Only three of the seven diagnostic tests may easily be applied to it. In order to permit nuclear exchange in the absence of cytoplasmic mixing it is necessary to separate the conjugating pair early in the mating process. The large size of the *Paramecium* cell permits the introduction of either mitochondria or killer particles by the process of microinjection. This allows these genomes to be exposed to mutagens *in vitro* quite independently of the nuclei. These techniques are very powerful ones and confirm the continued usefulness of this organism in spite of its disadvantages.

Fungi The fungi have proved to be the most useful microbial eukaryotes for genetic studies. They may be grown on defined media on simple carbon substrates and are also sexually very versatile, permitting a wide range of genetic techniques to be employed. In terms of the experimental techniques which may be applied to them the fungi may be divided into the yeasts and the filamentous fungi. The yeast *Saccharomyces cerevisiae* and the filamentous ascomycete *Neurospora crassa* (the pink bread mould) have both been extensively used in studies on extrachromosomal inheritance and their life cycles will be examined as representatives of these two groups of fungi.

Saccharomyces cerevisiae, the bread and ale yeast, is a unicellular ascomycete which divides by budding. The unicellular habit of yeast means that it may be handled exactly like bacteria and this commends it as an experimental subject. However, the principal advantage of *S. cerevisiae* in the context of mitochondrial genetics is that, unlike other fungi, it is not an obligate aerobe but may grow and divide using an entirely fermentative metabolism, dispensing with the functions of oxidative metabolism supplied by the mitochondria. This means that mutations in mitochondrial genes are not lethal. Indeed, the organism can still grow and divide on a fermentable substrate even when the entire mitochondrial genome has been lost.

The life cycle of *S. cerevisiae* is described in Figure 4. The organism can grow and divide in either the haploid or the diploid form permitting the identification of nuclear mutants in haploids and the definition of functional genes by complementation testing in the diploids. If the diploid form is subjected to nitrogen starvation it will undergo meiosis to produce four haploid ascospores. These four spores may be separated by microdissection and this permits tetrad analysis to be carried out. Alternate alleles of a given nuclear gene, for instance the a and α forms of the mating-type gene, segregate 2:2 in the meiotic tetrad, i.e. $2a{:}2\alpha$, whilst extrachromosomal genes segregate 4:0.

The diploid state is reconstituted by the fusion of an a cell with an α cell. Nuclear fusion (karyogamy) generally follows immediately after cytoplasmic fusion (plasmogamy) and thus there is no stable heterokaryon formed to allow the distinction between nuclear and cytoplasmic genes. However, in some strains, nuclear fusion is delayed and this permits the isolation by micromanipulation of haploid buds containing a mixed cytoplasm. This heterokaryon test is the only one of the seven diagnostic tests which presents any problems in *Saccharomyces*. Yeast is therefore a near perfect system for the study of mitochondrial genetics. Its utility may be compared to that of other organisms by reference to Table 3.

Neurospora crassa, like *S. cerevisiae*, is an ascomycete fungus which produces haploid ascospores (eight in this case) within a sac called an ascus. The ascus of *Neurospora* is elongated and the linear arrangement of the eight spores within it reflects the relationship between sister chromatids at meiosis I. Such an ordered

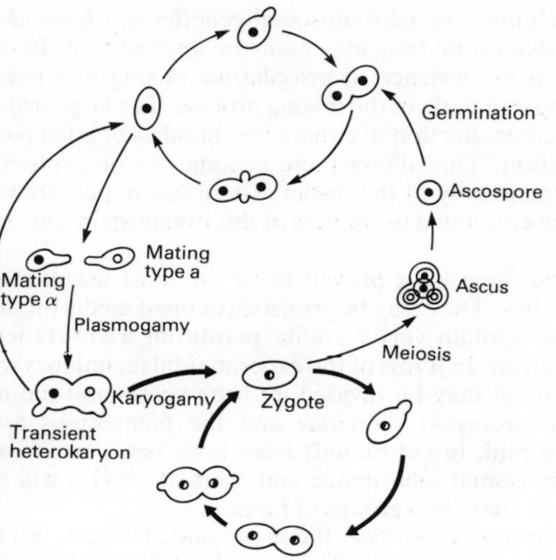

Fig. 4 Life cycle of *Saccharomyces cerevisiae.*

tetrad enables the investigator to distinguish between segregations which occur at the first and second divisions of meiosis. This has made *Neurospora* an extremely useful tool for the study of genetic recombination.

The main advantage of *Neurospora* (and other filamentous ascomycetes such as *Aspergillus nidulans*) in the study of extrachromosomal genetics is the formation of a stable heterokaryon. It will be recalled that a heterokaryon is a form in which the haploid nuclei from each parent strain coexist (without fusion) within the same cytoplasm. This stage in the fungal life cycle permits the distinction to be made between an extrachromosomal gene within the nucleus and a true cytoplasmic gene.

A diagram of the life cycle of *Neurospora crassa* is shown in Figure 5. In *S. cerevisiae* the heterokaryon stage was transient and the diploid stage persisted; in *Neurospora* the reverse is true. Aerial hyphae from the heterokaryotic vegetative mycelium may produce either macroconidia or microconidia. The macroconidia contain a number of nuclei and re-establsh the heterokaryon when they germinate. The microconidia, on the other hand, are uninucleate and are able to fuse with a specialized hypha (the trichogyne) of complementary mating type. The trichogyne is borne on a structure called the protoperithecium which is the precursor to the fruiting body or perithecium, the latter containing the mature asci. The fusion of two parental nuclei within the perithecium initiates the differentiation of an ascus. Meiosis begins immediately after nuclear fusion and yields eight haploid ascospores. Within this group of eight spores, still called a meiotic tetrad, chromosomal genes segregate 4:4 whereas extrachromosomal genes segregate 8:0. The ascospores germinate to produce haploid mycelia which can form heterokaryons through hyphal fusion with mycelia of oppositing mating type.

All seven diagnostic tests for extrachromosomal genes may be applied to *Neurospora* (see Table 3). Its nuclear genetics has been extensively studied; it

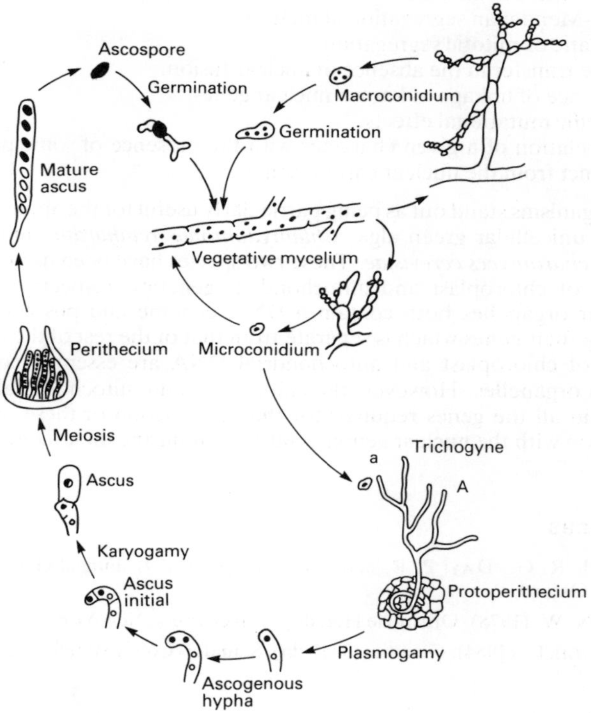

Fig. 5 Life cycle of *Neurospora crassa.*

forms stable heterokaryons and it is possible to perform microinjection on it. In spite of all these advantages, however, *Neurospora* is a much less popular tool for the study of extrachromosomal genetics than *S. cerevisiae*. There are two main reasons for this. The first is that unicellular organisms are easier to handle than filamentous ones and, more importantly, *Neurospora* is an obligate aerobe and this severely limits the type of mitochondrial mutations that can be isolated.

Summary

Extrachromosomal or non-Mendelian genetics is the study of the behaviour of those genes which do not occur on nuclear chromosomes. It is mainly concerned with the genetics of cytoplasmic organelles but a number of other extra-chromosomal genes are known, some of which have been characterized. Micro-organisms have been found to be useful subjects for the study of this kind of genetics. Their metabolic and sexual versatility permits the application of a number of diagnostic tests which identify the extrachromosomal location of a given gene. Seven such tests have been discussed:

1. Differences in inheritance patterns between reciprocal crosses.
2. Non-Mendelian segregation at meiosis.
3. Somatic or mitotic segregation.
4. Gene transfer in the absence of nuclear fusion.
5. Absence of linkage to known nuclear genes.
6. Specific mutational effects.
7. Correlation of a given character with the presence of some physical entity distinct from the nuclear chromosomes.

Two organisms stand out as being particularly useful for the application of these tests; the unicellular green alga, *Chlamydomonas reinhardtii*, and the budding yeast, *Saccharomyces cerevisiae*. These two species have been used extensively in the study of chloroplast and mitochondrial genetics, respectively. These two subcellular organelles both contain a DNA genome and possess a system for expressing their genes which is separate from that of the rest of the cell. The gene products of chloroplast and mitochondrial DNA are essential components of those two organelles. However, the chloroplast and mitochondrial genomes do not contain all the genes required for the construction of these organelles and cooperation with the nuclear genome must occur during their biogenesis.

References

FINCHAM, J. R. G., DAY, P. R. and RADFORD, A. (1979). Fungal Genetics (4th Edn). Blackwell, Oxford.

GILLHAM, N. W. (1978). Organelle Heredity. Raven Press, New York.

GOODENOUGH, U. (1984). Genetics (3rd Edn). Saunders College Publishing, Philadelphia.

2 Mitochondrial genetics

There have been three important phases in the development of mitochondrial genetics. Classical genetic studies suggested the existence of a mitochondrial genome and this was subsequently confirmed by biochemical studies. This understanding of the molecular basis of mitochondrial genetics has permitted the construction of a detailed map of the mitochondrial genome and an analysis of the process of mitochondrial recombination. All of these studies have centred on *Neurospora crassa* and *Saccharomyces cerevisiae* as these organisms permit the application of the widest range of genetic techniques. The life cycles of these organisms are illustrated in Figures 4 and 5 and reference to these will aid the understanding of the experiments described in this chapter.

Poky and petite—the discovery of mitochondrial genetics

The yeast *Saccharomyces cerevisiae* is a facultative anaerobe and can therefore dispense with the capacity to carry out oxidative phosphorylation and electron transfer. This means that mutants which have suffered the loss or severe derangement of their mitochondrial genome may be isolated in this organism whereas mutants obtainable in obligate aerobes such as *Neurospora crassa* and *Paramecium aurelia* are necessarily less severe in their effect. It is impossible, however, for yeast to dispense with mitochondria all together. The mitochondrial genome is concerned solely with the maintenance of the respiratory functions of the organelle. However, mitochondria carry out a wide range of essential functions. In addition to the role of the inner mitochondrial membrane in electron transfer and oxidative phosphorylation, it also contains enzymes concerned with fatty acid oxidation and lipid biosynthesis, and the matrix contains the enzymes of the tricarboxylic acid cycle as well as some enzymes concerned with amino acid metabolism.

The cytoplasmic petite mutation in yeast

This mutation was discovered by Boris Ephrussi in the late 1940s. He noticed that when yeasts were grown on glucose agar about 1% of the colonies were unusually small. This character, which he termed *petite colonie*, was found to be due to respiratory incompetence of the cells. The petite cells could only ferment glucose but not respire it. Since fermentation is less efficient than respiration, the petite colonies grew more slowly than the wild type which he called *grandes*.

The petite mutation is very unusual because it occurs at extraordinarily high frequencies. Laboratory strains of yeast carrying 10% petite mutants are not unusual and a strain carrying 50% has been found. These frequencies should be compared to 0.001% for spontaneous mutations of known nuclear genes in the same organism. In addition, the petite mutation occurs with undiminished frequency not only in haploid cell lines but in diploids and tetraploids as well and, unlike point mutations of nuclear genes, the cytoplasmic petite mutation never reverts to wild type. Another distinctive feature of the cytoplasmic petite mutation is that it can be induced at frequencies up to 100% by chemical mutagens such as acriflavine, ethidium bromide and 5-fluorouracil which are unable to induce mutations in any known nuclear gene. Specific mutagenesis of this type has already been noted as a diagnostic feature of non-Mendelian genes.

Ephrussi recognized that the petite phenotype was due to the cells being incapable of respiration, since they cannot grow on the non-fermentable substrates such as glycerol. This phenotype is not the result of the loss of any single enzyme concerned with respiratory function but is due to the loss of many such enzymes. Thus, the petite mutation is pleiotropic in its effect; petite cells lacking a number of mitochondrial enzymes including cytochromes a and b (but not c), succinate dehydrogenase, cytochrome c oxidase, NADH-cytochrome c oxidoreductase and α-glycerolphosphate dehydrogenase.

Ephrussi and his colleagues studied the inheritance of this unusual phenotype and found that when a petite and a grande (wild type) strain were crossed, the resulting zygote had a grande phenotype. If that zygote, or the diploid descendants of it, were sporulated the four spores of the ascus were all grande (Figure 6). Such a 4:0 segregation in meiosis is diagnostic, but not definitive, evidence for non-Mendelian inheritance. It is possible that the petite phenotype resulted from the simultaneous mutation of a number of unlinked nuclear genes to their recessive alleles. This was checked by performing repeated backcrosses of haploids derived from a petite × grande zygote to the petite parent. The petite character did not reappear after five successive backcrosses and it was calculated that the simultaneous mutation of 16 unlinked genes would have had to occur to account for this result. This is theoretically possible as the yeast nuclear genome is arranged in 17 linkage groups, but it is not reconcilable with the extraordinarily

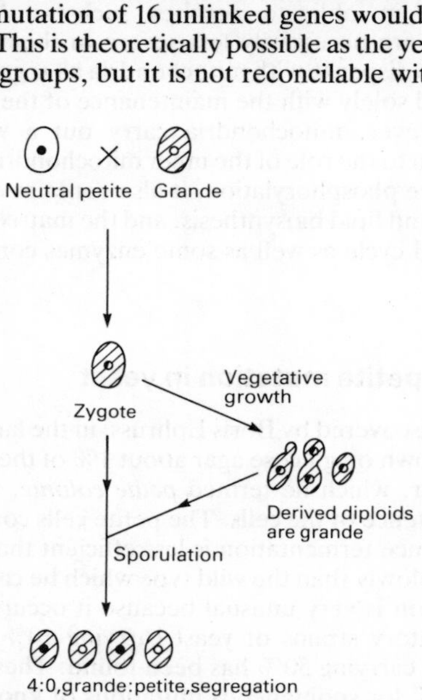

Fig. 6 The neutral petite x grande cross.

high spontaneous frequency of the petite mutation. This is convincing evidence of an extrachromosomal mode of gene transmission but it satisfies only one of the diagnostic tests described in Chapter 1.

The further genetic characterization of the petite mutation was aided by the discovery that there were two types of petite mutant. One, the neutral petite, has already been described in Figure 6; the second type is known as the suppressive petite and is described in Figure 7. Suppressive petites, when crossed with grandes, produce zygotes which, if sporulated immediately after their formation, give rise to four petite spores in the meiotic tetrad. If the zygote is not immediately sporulated but allowed to divide mitotically, the diploids produced are a mixture of grandes and petites. The higher the proportion of petites in the diploid clone the more 'suppressive' the mutation is said to be. It has already been noted that mitotic segregation of this sort is a diagnostic feature of extrachromosomal inheritance.

The discovery of the suppressive petites allowed Wright and Lederberg to apply the heterokaryon test and demonstrate that the petite mutation was not only extrachromosomal but truly cytoplasmic. Their experiment is shown in Figure 8. These workers exploited the fact that in a certain variety of wine yeast the heterokaryon stage persists for a reasonable time. They crossed a suppressive petite with a grande strain and removed the haploid buds produced by the mitotic division of the two parental nuclei in the heterokaryon. They allowed these buds to grow into colonies and were able to show that the petite character segregated independently of its parental nucleus. This demonstration was only possible because genetic markers which identified the two parental nuclei were available. This emphasizes the importance of studying extrachromosomal genetics in an organism in which chromosomal genetics is well developed.

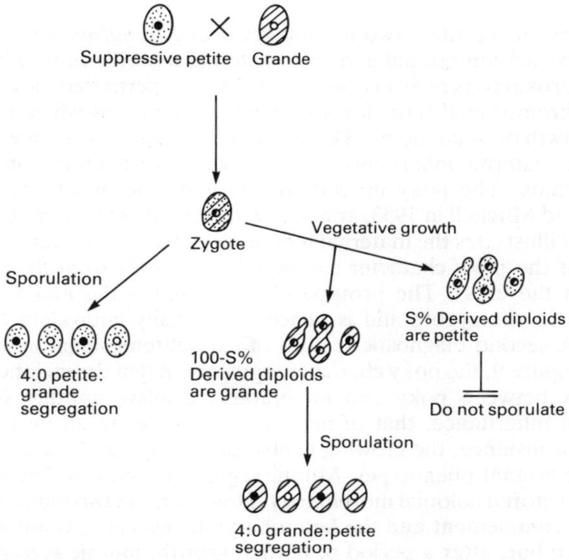

Fig. 7 The suppressive petite x grande cross.

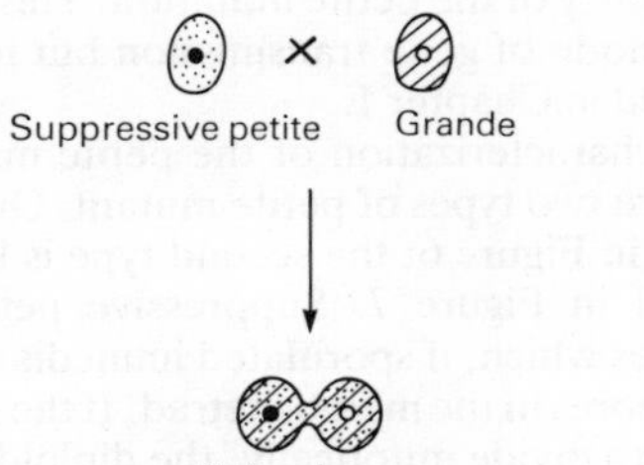

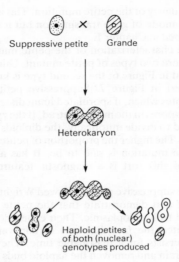

Fig. 8 The heterokaryon test.

The molecular basis of the cytoplasmic petite mutation and the phenomenon of suppressiveness will be discussed later in the chapter.

Poky and mi mutants in *Neurospora*

Neurospora crassa offers two advantages over *Saccharomyces cerevisiae* for the study of extrachromosomal genetics: maternal inheritance may be detected and stable heterokaryons can be constructed. This has permitted the demonstration of an extrachromosomal basis for a number of mutations which result in slow or erratic growth on solid media. These mutants include 'poky', one of the *mi* series, named for *m*aternal *i*nheritance, the SG (slow growth) series and the *stp* (stop–start) mutants. The poky mutants (*mi*-1) were the first to be discovered, by Mitchell and Mitchell in 1953, and they will be used as an example of the type.

Figure 9 illustrates the maternal inheritance of the *mi*-1 mutation. The diagram shows that the poky character can be inherited only from the protoperithecial partner in the cross. The protoperithecium makes the major contribution of cytoplasm to the zygote and is hence functionally equivalent to the maternal partner. A second diagnostic feature of extrachromosomal inheritance is illustrated in Figure 9, the poky character segregates 8:0 in the meiotic tetrad.

A cross between poky and *mi* mutants displays another feature of non-Mendelian inheritance, that of mitotic segregation. In an *mi*-1 × *mi*-3 heterokaryon, for instance, the growing hyphal tips always carry determinants for only one of the mutant phenotypes. Mitotic segregation here is very rapid and shows itself in a sectored colonial morphology. However, the two mutants *mi*-1 and *mi*-4 appear to complement and the heterokaryotic mycelium is initially wild type in appearance but, after a period of normal growth, mitotic segregation produces sectors of one or the other mutant type.

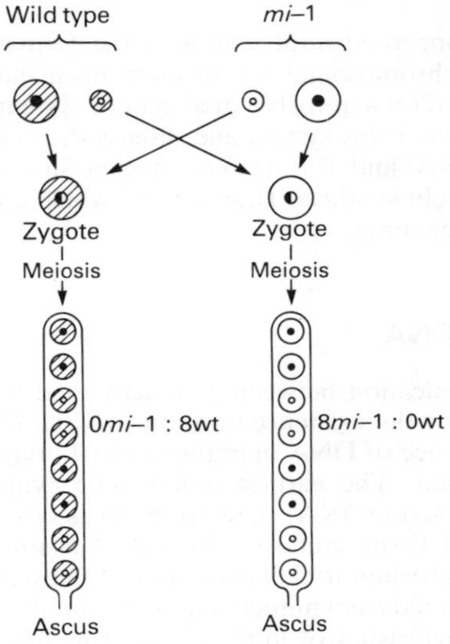

Fig. 9 Maternal inheritance of the poky (*mi*-1) mutation in *Neurospora crassa.*

Srb, in studying the SG mutants of *Neurospora*, has elegantly demonstrated their cytoplasmic determination. He made interspecific crosses between an SG mutant of *N. crassa* and wild type *N. sitophila*. In the first cross *N. crassa* was the protoperithecial parent and *N. sitophila* the conidial one. *N. sitophila* then remained the conidial parent for a series of successive backcrosses. In this way the cytoplasm of the SG *N. crassa* parent is maintained whilst the nuclear genome eventually becomes identical with that of the *N. sitophila* parent. Since the SG character was maintained throughout this series of crosses its genetic determinant must be located in the cytoplasm.

Maternally-inherited genes in *Neurospora*, including poky, are similar to the cytoplasmic petite mutation in *Saccharomyces* in that they impair the respiratory function of the mitochondria due to a number of effects rather than a single mutational change. Poky mutants are deficient in both cytochromes a and b but contain an unusually high amount of cytochrome c. The cytochrome complement of some of the SG and *stp* mutants is similar to that of poky. The rest of the *mi* series are deficient in cytochrome a but have almost normal amounts of cytochromes b and c. The effect of these pleiotropic mutations is not as severe as in the petites of yeast. Indeed, older cultures appear to recover from the effects of these mutations and grow at rates which approach those of the wild type. This has led some authors to describe this series of mutations in *Neurospora* as developmental defects: the normal development of a fully functional mitochondrion appears to be slowed down. A knowledge of the molecular basis of these kinds of mutations and how they influence mitochondrial function will help us to understand these phenomena.

The molecular biology of the mitochondrion

We have been concerned until now with the formal genetic evidence for the existence of extrachromosomal genetic elements in eukaryotic organisms and, in particular, with that for a mitochondrial genetic system. It is now time to examine the molecular basis of this system and discuss the evidence for the existence of mitochondrial DNA and for distinct mechanisms of RNA synthesis (transcription) and protein synthesis (translation) which are used to give phenotypic expression to this genome.

Mitochondrial DNA

A lack of communication between geneticists and biochemists meant that the genetic evidence for the existence of mitochondrial DNA was ignored for some time and the presence of DNA in mitochondrial preparations was regarded as a nuclear contaminant. The earliest non-genetic evidence for the existence of mitochondrial DNA (mitDNA) came from the electron microscopic examination of thin sections of tissue culture cells and of *Neurospora* which revealed the presence of intramitochondrial fibrils of DNA. Subsequently a distinct species of DNA has been isolated from mitochondria of a number of organisms. Table 4 lists some of the characteristics of mitDNA for a number of genetically important species. The size of the mitDNA of most of these microorganisms falls within the range of 15 to 25 μm in contour length with *C. reinhardtii* being an exception at only 4 to 6 μm. While this value is much larger than that for mammalian

Table 4 Characteristics of mitDNA from genetically important species

Species	Size (μm)	Buoyant density (g/cc)	%G + C	Conformation
Algae				
Chlamydomonas reinhardtii	4–6	1.706	48	circular?
Euglena gracilis		1.688–1.691	31	linear?
Protozoa				
Paramecium aurelia	14–17.5	1.702	40	hair-pin loop
Slime moulds				
Dictyostelium discoideum		1.688	28	linear?
True fungi				
Saccharomyces cerevisiae	21–25	1.683	17	circular
Neurospora crassa	19–25		40	circular
Aspergillus nidulans	11			circular
Podospora anserina	31	1.694		circular

mitDNAs, which are circles of about 5 μm in circumference, the amount of useful information contained by these molecules is not necessarily proportionally greater. For instance, detailed studies of the structure of *S. cerevisiae* mitDNA by Bernardi and his co-workers have revealed that half of the molecule consists of spacer regions rich in adenosine and thymidine which are not expressed as protein products.

Replication of mitDNA This process has been studied most intensively using mammalian mitDNA where the much smaller physical size of the molecule means that the experiments present far fewer technical problems. The method of replication elucidated by these studies has since been found to apply also to the smaller bacterial plasmids.

The covalently closed, circular mitDNA molecule is replicated by a single replication fork travelling in one direction, whereas the replication of both prokaryotic and eukaryotic chromosomes is bidirectional. A replication fork is the structure formed by the separation of the two parental DNA strands as the daughter strands are synthesized. DNA polymerase enzymes can only synthesize DNA in the 5′ to 3′ direction and therefore only one daughter strand may be synthesized continuously. Although these fragments are individually synthesized in the 5′ to 3′ direction the overall direction of synthesis of the whole strand is 3′ to 5′ (see Figure 10a). If there is a single replication fork there is only one overall direction of synthesis (unidirectional replication). In bidirectional replication, on the other hand, there are two forks which migrate in opposite directions, away from one another (Figure 10b).

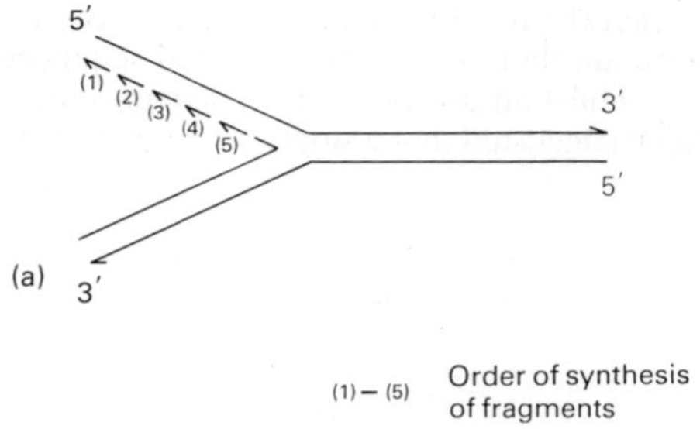

Fig. 10a A DNA replication fork.

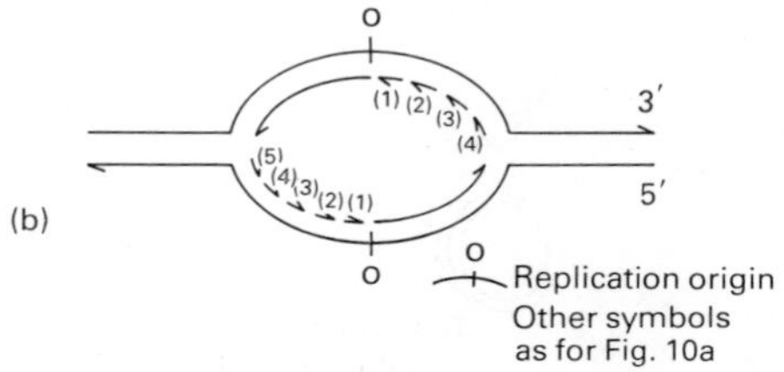

Fig. 10b Bidirectional replication.

Replication of the circular mitDNA is initiated by the synthesis of an oligonucleotide, called the E-strand, which is complementary to part of one parental strand. The other parental strand is displaced in this region to form a structure called the D-loop. The E-strand and the D-loop are extended for some distance before nicks are introduced into the parental strands and the complement of the D-loop strand begins to be synthesized (Figure 11a). Such data as are available on the replication of the larger circular DNA molecules of the lower eukaryotes are compatible with a D-loop mechanism but this has yet to be demonstrated convincingly.

In some protozoans the mitDNA is linear and more information is available on its mechanism of replication, and both unidirectional and bidirectional modes have been discovered in different organisms. The replication of a linear DNA molecule presents special problems. This is because DNA polymerases are only able to add nucleotides to the free 3′ hydroxyl group of a pre-existing polydeoxyribonucleotide or polyribonucleotide chain. These enzymes are unable to initiate the synthesis of a DNA strand *de novo*. This property of polymerase enzymes does not matter in the replication of circular molecules since there are no free ends to the template strands. However, in linear molecules it is impossible to replicate the DNA near the two free ends. In the two protozoans which have been examined this problem is solved in different ways, but both involve the arrangement of the terminal portion of the molecule into a looped structure which may be replicated.

In *Tetrahymena pyriformis* a bidirectional mechanism is employed. Replication is initiated near the middle of the molecule and the two replication forks migrate outwards, one towards each end. Results obtained in Borst's laboratory have suggested that the problem of replicating the ends of the linear mitDNA molecule is solved in *Tetrahymena* by having a number of sequences near its ends which are direct repeats of one another. A directly repeated sequence is one of the type ABC . . . ABC . . . whilst an inverted repeat sequence has the form ABC . . . . . . CBA. It has been suggested that a single-strand nick is made at a specific site

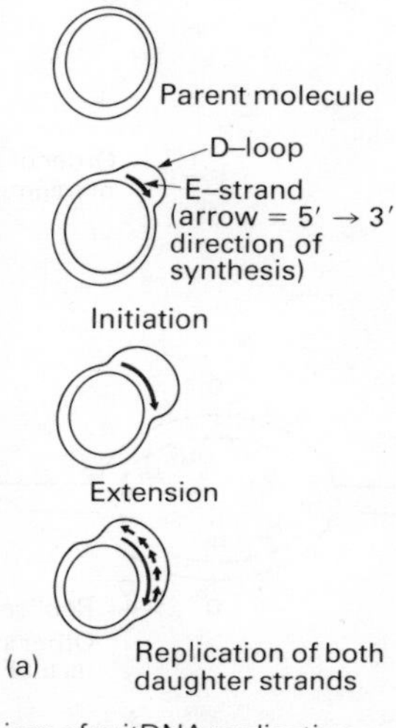

Fig. 11a The D-loop mechanism of mitDNA replication.

adjacent to the repeated sequences. DNA synthesis from this nick would then displace a single strand which is available to base pair with the unreplicated terminus of the molecule, forming a loop. DNA synthesis through the loop would then serve to replicate the end of the molecule and endonucleolytic cleavage at the original site, followed by ligation of the nicks, would regenerate a linear molecule. Figure 11b should make this rather complicated scheme easier to understand.

Paramecium aurelia mitDNA is replicated in a unidirectional manner. In this case the problem of replicating the ends of the molecule is solved by converting the entire molecule into a hairpin loop by ligating the 5′ and 3′ terminal nucleotide at one end. Replication begins very close to that end, forming a replication bubble which produces a banjo-shaped molecule. The replication fork proceeds in one direction, enlarging the bubble, until the entire molecule has been replicated. The product of replication by this mechanism is a linear dimer which must be cut by a double strand endonuclease to yield the two daughter DNA molecules. The scheme is illustrated in Figure 11c.

Mitochondrial rRNA Most mitochondrial rRNAs have a much lower guanosine and cytosine content than their cytoplasmic counterparts encoded by nuclear DNA. This results in a more flexible secondary structure and gives them anomalous electrophoretic properties. Reliable estimates of the molecular weight of mit rRNA therefore may only be made using gel electrophoresis under strongly denaturing conditions or by electron microscopy where the monolayers have been prepared on a denaturing hypophase. Table 5 lists the characteristics of the mit

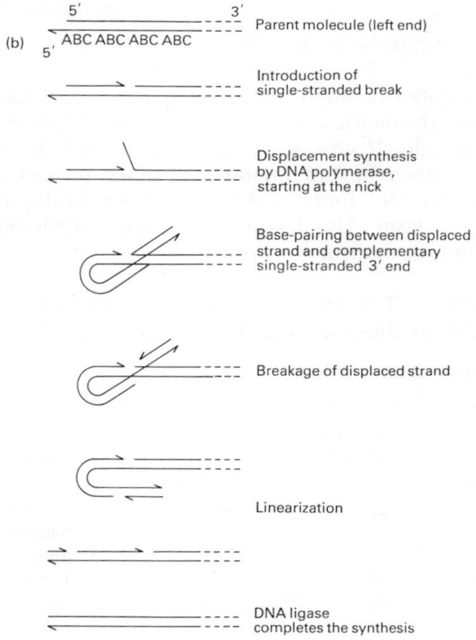

Fig. 11b Replication of *Tetrahymena* mitDNA.

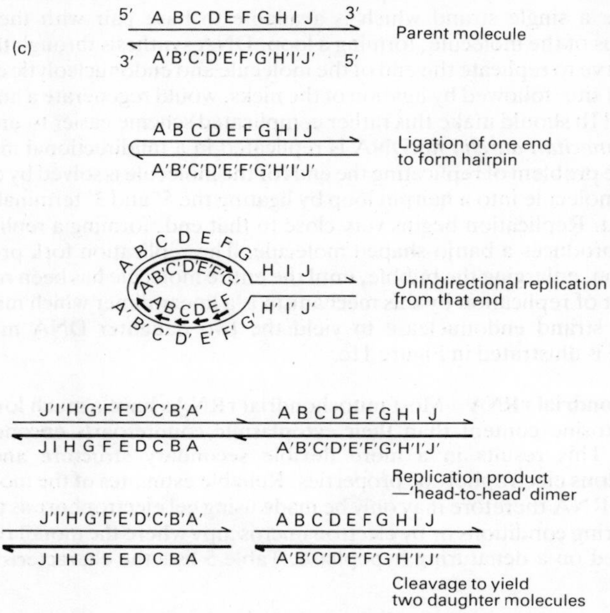

Fig. 11c Replication of *Paramecium* mitDNA.

rRNAs from a number of important species. Another unusual feature of mit rRNAs is that they are much less methylated than the cytoplasmic species. Indeed, some mammalian mit rRNAs are completely unmethylated. In addition, it appears that in *S. cerevisiae* and a number of other microorganisms there is no equivalent of the 5S rRNA found in the ribosomes of both prokaryotic cells and the eukaryotic cytoplasm. This feature makes the mitochondrial ribosomes of these organisms unique in the living world.

Mitochondrial tRNAs The tRNAs encoded by mitDNA have fewer methylated and unusual bases than those specified by nuclear DNA. The number of different

Table 5 Mitochondrial ribosomal RNAs

Species	S-values of molecules		% G + C	Methylation
Saccharomyces cerevisiae	22	15	48	2 methylated bases in 22S, none in 15S
Neurospora crassa	24	17	35–38	1–3%
Aspergillus nidulans	23	16	27–32	—
Euglena gracilis	21	16	30	—

mit tRNAs in any one species is quite low, about 26, and it is possible that some cytoplasmic tRNAs may be recruited into the mitochondria. Isoaccepting species of tRNA, that is molecules that are charged with the same amino acid but which differ in their anticodons (e.g. $tRNA^{Glu}_{GAA}$ and $tRNA^{Glu}_{GAG}$) have been detected in yeast mitochondria but do not appear to be common. It is becoming evident that the mitochondria of the lower eukaryotes use only a restricted range of codons to specify the amino acids in proteins. Those in yeast, for instance, have a bias towards codons containing adenosine and uridine. It has recently been found that mitochondria have a slightly different genetic code to that used by bacteria or eukaryotic nuclei. This conclusion has been reached by comparing the amino acid sequence of a protein with the nucleotide sequence of the gene which specifies it. This point will be developed further in the section on genetic mapping.

Mitochondrial mRNAs Messenger RNA molecules synthesized in the nucleus have distinct terminal structures. The 5′ end is 'capped' by a methylated guanosine molecule and the 3′ end carries a polyadenylic acid 'tail'. Both of the structures are added after transcription. Mitochondrial mRNA does not have the methylated guanosine cap and the question of whether it carries a poly (A) tail has provoked some controversy. However, studies with yeast by Rabinowitz' group in Chicago have made it clear that at least a proportion of the yeast mitochondrial messages carry a poly (A) tail. A number of studies have linked the shortening of poly (A) tails on mRNAs to their functional half-life. In *Saccharomyces* the average half-life of mit mRNA is less than 10 minutes, approximately half that of the average cytoplasmic mRNA. This may explain why this tail is so difficult to detect and why only a small proportion of the molecules appear to carry it.

Mitochondrial RNA polymerase

Mitochondria appear to contain the simplest RNA polymerases yet discovered. Instead of the large multi-subunit enzyme complexes which are found in both eukaryotic nuclei and in prokaryotes, the mitochondrial RNA polymerases of both *Neurospora crassa* and *Saccharomyces cerevisiae* are simple polypeptide chains of 64,000 and 68,000 molecular weight respectively.

The drug rifampicin is an inhibitor of the initiation of RNA synthesis in bacteria but has no effect on nuclear RNA synthesis in eukaryotes. Much research effort has been expended on the question of the rifampicin sensitivity of mitochondrial RNA polymerase and conflicting results have been obtained. Difficulties arise because most eukaryotic microorganisms do not take up this inhibitor and so experiments have to be performed on cell-free extracts, isolated mitochondria, submitochondrial particles or purified polymerases. Use of the first three techniques requires the drug to cross the mitochondrial membrane, the permeability of which may vary in different preparations. Studies with purified polymerases do not suffer this drawback and here the weight of evidence suggests that the enzymes are rifampicin sensitive.

A number of other inhibitors preferentially inhibit mitochondrial RNA synthesis and may be used *in vivo*. These include ethidium bromide, acridine dyes, daunomycin and cordycepin (3′-deoxyadenosine). The first three act by binding preferentially to mitDNA. It is not clear whether the inhibitory effect of cordycepin is due to premature termination of RNA chains or is an indirect result of an inhibition of ATP metabolism. A number of these inhibitors have proved useful experimentally, for instance, in the determination of messenger half-life and in the elucidation of the mode of action of some petite mutagens.

It has been demonstrated in mammalian mitochondria that transcription of the DNA is a symmetrical process with most of one of the two complementary RNA strands subsequently being degraded. For the microbial eukaryotes there is no firm evidence that transcription of mitDNA is other than asymmetric. If symmetrical transcription does occur in this group of organisms the non-sense RNA product must be very rapidly degraded since it cannot be detected experimentally.

Mitochondrial protein synthesis

The initial demonstration that mitochondria have their own protein synthesizing system was made with isolated mammalian mitochondria. Such *in vitro* techniques have been less useful with microorganisms due to the difficulty of preparing intact mitochondria from cells with a tough outer wall. However, studies on microbial systems have been assisted by the fact that mitochondrial translation shows a number of similarities with prokaryotic protein synthesis and is quite distinct from that occurring in the eukaryotic cytoplasm. At the same time, this meant that early workers using *in vitro* systems had rigorously to demonstrate that their results were not due to bacterial contamination.

Mitochondrial protein synthesis is sensitive to drugs such as chloramphenicol, erythromycin, spiramycin, lincomycin, oleandomycin and mikamycin, all of which also inhibit bacterial protein synthesis, but is resistant to cycloheximide, a drug which inhibits protein synthesis by the cytoplasmic ribosomes of eukaryotes. These results stimulated the idea that mitochondrial ribosomes were very similar to bacterial ribosomes and much has been made of the fact that, in many experiments, values close to 70S have been found for the sedimentation coefficient of mitochondrial ribosomes. Bacterial ribosomes also have an S-value of 70S whereas eukaryotic cytoplasmic ribosomes are 80S. These comparisons should, however, be treated with some caution. Values between 72S and 80S have been published for the mitochondrial ribosome of *S. cerevisiae*.

The S-value of such a large complex as a ribosome is dependent on both its shape and the conditions under which centrifugation or electrophoresis is performed. Some mitochondrial ribosomes have a very unusual structure, for instance, those of *Tetrahymena pyriformis* are composed of two subunits of almost equal size. Moreover, in many microorganisms the mitochondrial ribosomes appear intimately associated with the inner mitochondrial membrane and there may be difficulties in separating the ribosome particles from membrane components. Finally, it should be emphasized once more that the lack of a 5S ribosomal RNA makes microbial mitochondrial ribosomes unique.

The differential effect of various protein synthesis inhibitors on cytoplasmic and mitochondrial ribosomes has permitted the use of *in vivo* labelling experiments to demonstrate which proteins found in mitochondria are synthesized by each translation system. Experiments in which cytoplasmic protein synthesis is inhibited by cycloheximide, and radioactive amino acids are used to label proteins synthesized by the mitochondrial system, have been particularly popular. In *Neurospora crassa* six mitochondrial proteins were labelled under these conditions; they have molecular weights of 33,500, 27,700, 25,000, 21,000, 17,500, and 11,000. In *Aspergillus nidulans* four mitochondrial membrane proteins of molecular weights 40,000, 27,000, 18,000, and 13,000 showed cycloheximide-resistant labelling. In *S. cerevisiae* five membrane proteins (molecular weights 48,000, 33,00, 28,000, 23,000 and 15,800) were labelled in the presence of cycloheximide but not when mitochondrial transcription or translation was inhibited by use of drugs or temperature-sensitive mutants. These data demonstrate that mitochondrial DNA and the systems required for its transcription and translation exist to produce a small number of proteins of comparatively low molecular weight which are located within the inner mitochondrial membrane.

There is a problem with these inhibitor studies. The construction of a functional mitochondrion is a cooperative effort requiring the gene products of both nuclear and mitochondrial DNA. If protein synthesis by one system is prevented one may be removing some component which is essential for the synthesis of certain proteins made by the other system or may prevent their incorporation into the mitochondrial membrane. There are a number of instances of such cooperative effects but an example drawn from Tzagaloff's work on the yeast ATPase complex, will illustrate the point.

Glucose derepression brings about an increase in the oligomycin-sensitive ATPase activity in the mitochondria. This increase in enzyme activity may be prevented by inhibiting mitochondrial protein synthesis with chloramphenicol. The straightforward conclusion is that the ATPase is synthesized in the mitochondria, but this is not the case. Under conditions of glucose derepression and chloramphenicol inhibition, a soluble ATPase with biochemical and immunological properties identical to those of the mitochondrial F1 complex accumulates in the cytoplasm. A protein identical to the oligomycin-sensitivity conferring protein (OSCP) which links the F1 ATPase to the inner mitochondrial membrane also accumulates in the cytoplasm. The accumulation, under these conditions, of both these components of the mitochondrial ATPase complex may be prevented by the inhibition of cytoplasmic protein synthesis with cycloheximide and it therefore appears that the F1 ATPase and its OSCP are synthesized on cytoplasmic ribosomes. The results obtained with chloramphenicol indicate, however, that some products of the mitochondrial translation system are required for the attachment of these components to the inner mitochondrial membrane.

It is likely that all of these proteins are membrane-bound and one has been identified as the DCCD-binding lipoprotein, the so-called subunit 9 of the intact ATPase complex. A full discussion of all the mitochondrial components encoded by mitochondrial DNA will be deferred until the detailed treatment of mitochondrial genetics has been completed.

A molecular explanation of the poky and petite mutations

The discussion on the cooperation of the nuclear and mitochondrial genomes in the synthesis of the ATPase complex and its integration into the inner mitochondrial membrane shows how mitochondrial mutations can have pleiotropic effects. Although only a few proteins are synthesized within the mitochondrion they are responsible for the correct integration into the organelle of a much larger number of proteins which are encoded in the nucleus and synthesized in the cytoplasm. Any mutation which prevents mitochondrial protein synthesis will prohibit the functioning of a wide range of respiratory enzymes. Both the poky mutation in *N. crassa* and the petite mutation in *S. cerevisiae* prevent, in quite different ways, the process of mitochondrial translation.

The mitochondrial DNA of *S. cerevisiae* bands at a distinct equilibrium density in a CsCl gradient (see Table 4). If the DNA from cytoplasmic petite mutants is examined in a similar manner it is found that the mitDNA either bands at a density which is markedly different from that of the grande parent (it is usually less dense) or it is not detectable at all. It is evident that the cytoplasmic petite mutation results in the loss or severe derangement of mitDNA. Petites containing no mitDNA at all are referred to as ρ^0 petites and those containing altered DNA as ρ^-; grande cells are designated ρ^+. The mitDNAs of ρ^- mutants have been examined by both hybridization and restriction enzyme analysis and it has been found that while some large segments of the genome have been deleted other segments have been reiterated many times.

Petite cells of either type, ρ^- or ρ^0, are incapable of carrying out mitochondrial transcription or translation. Since it is known that mitochondrially synthesized proteins are essential to the integration of a wide range of respiratory enzymes into the organelle membrane, the pleiotropic nature of the petite mutation is explained. It also follows that all cytoplasmic petites have exactly the same phenotype although the nature of the change in mitDNA which they have undergone differs widely. This contrasts with the nuclear or segregational petite mutants where a single mutation in a chromosomal gene results in the loss of the activity of just one respiratory enzyme.

Two classes of cytoplasmic petite mutant were recognized genetically, neutral petites and suppressive petites. The molecular explanation for this genetic difference appears to be that the mitochondrial DNA of suppressive petites contains a high number of replication origins relative to its size and can replicate faster than grande mitochondrial DNA. This results in the petite DNA dominating the cytoplasm in zygotes produced from a grande × suppressive petite cross. The progeny cells of these zygotes are therefore more likely to inherit a petite than a grande mitochondrial genome. Neutral petites, on the other hand, either lack mitochondrial DNA altogether (are ρ^0) or contain a mitochondrial genome which is replicated inefficiently. Thus in a neutral petite × grande cross, it is the grande mitDNA which dominates the zygote's cytoplasm.

N. crassa is an obligate aerobe and since mitochondrial gene products are essential to respiratory metabolism, mutants in which mitDNA is either absent or drastically altered will not survive. No major difference between the mitDNAs of poky and wild-type *Neurospora* can be detected by CsCl density gradient centrifugation or by restriction analysis. The effect of the poky mutation, unlike the petite, is reversible; older cultures of poky mutants recover and grow at rates approaching those of the wild type. Nevertheless, the poky mutation is pleiotropic in its effect on respiratory enzymes and we would expect it to cause some disturbance in mitochondrial protein synthesis.

It has been shown that vegetative mycelia of poky mutants are deficient in small subunits of their mitochondrial ribosomes. This results in extremely low rates of mitochondrial protein synthesis and the consequent effect on respiratory metabolism means that the hyphal growth rate is reduced. As the mycelium approaches stationary phase the balance between the numbers of large and small subunits of the mitochondrial ribosomes is restored to some extent and the growth rate of the mycelium therefore increases.

Genetic analysis of mitochondrial DNA

The mitochondrial mutations which have been discussed have a pleiotropic effect and, in the case of the cytoplasmic petite mutation, involve drastic changes to the mitochondrial genome. A fuller understanding of the molecular biology of the mitochondrion led investigators to expect that point mutations affecting single functions should be obtainable. These mutants could then be used to study mitochondrial genetic recombination and to construct a map of the mitochondrial genome.

Drug-resistant mutants The discovery that mitochondria have a distinct protein synthesizing system, and that its products are essential to respiratory function, stimulated a search for mutants resistant to inhibitors of either bacterial protein synthesis or oxidative phosphorylation. It was expected that point mutations in the mitochondrial genome would produce such strains. Spontaneous mutants of this type are easy to select and have been obtained in a number of species including *Saccharomyces cerevisiae, Schizosaccharomyces pombe, Podospora anserina, Aspergillus nidulans* and *Paramecium aurelia*.

In bacteria, resistance to protein synthesis inhibitors often involves mutations which result in alterations to the constituent proteins of the ribosomes. The point mutations in the mitochondrial genome, however, do not result in altered ribosomal proteins: they map in the mitochondrial genes which encode rRNA. The mitochondrial genome contains very few genes which encode ribosomal proteins. One example is the protein specified by the *var*-1 gene of yeast mitochondrial DNA but this protein is not essential to ribosomal function. The analogous protein in the *Neurospora* mitochondrial ribosome is encoded in the nucleus. The nuclear genome is, therefore, the site of almost all of the genes coding for mitochondrial ribosomal proteins.

Mutants resistant to inhibitors of oxidative phosphorylation and electron transport, such as oligomycin and antimycin A, have been isolated in *S. cerevisiae, Schizosaccharomyces pombe* and *A. nidulans*. In *S. cerevisiae* the isolation of such mutants has permitted the mapping of the mitochondrial genes which encode components of the electron transport chain. These enzyme complexes are composed of proteins encoded by both the nuclear and the mitochondrial genomes and in *N. crassa* mutants resistant to these oxidative phosphorylation inhibitors have proved to be of nuclear, rather than mitochondrial, origin.

Mutants with specific defects in mitochondrial function An example of a mitochondrial mutant with a specific defect has already been discussed—the poky or *mi*-1 mutants of *N. crassa*. However, since the specific defects were in the assembly of the mitochondrial protein synthetic machinery, the mutations had pleiotropic effects. Another early example, the *acu*-10 mutant of the basidiomycete *Coprinus lagopus*, is an extrachromosomally transmitted mutation which prevents growth on acetate as sole carbon and energy source. Sixty mutants of *N. crassa* with a similar phenotype to *acu*-10 have been described but these defined seven nuclear genes which specify respiratory enzymes. The nuclear or segregational petite mutants of *S. cerevisiae* also define a series of nuclear genes which have quite specific effects on respiratory function.

The most fruitful search for mitochondrial mutants having specific effects on organelle function has been conducted with *S. cerevisiae* and has resulted in the isolation of the mit^- series of mutants. These are a series of respiratory deficient mutants which are distinguished from the cytoplasmic petites by the fact that they retain active mitochondrial protein synthesis. A large number of respiratory deficient mutants, identified by their inability to grow on the non-fermentable substrate glycerol, were isolated. Each of these mutants was then examined to determine whether it retained its mitochondrial protein synthesis system. This was done by measuring the incorporation of radioactive leucine into protein in the presence of cycloheximide, an inhibitor which prevents protein synthesis by cytoplasmic ribosomes. Petite cells are unable to incorporate leucine in the presence of the inhibitor whereas *mit*⁻ mutants retain this capacity. The *mit*⁻ series includes mutations in both regulatory and structural genes that control the synthesis of components of cytochrome oxidase, reduced ubiquinone cytochrome c reductase and the oligomycin-sensitive ATPase complex.

Temperature-conditional mutants Mutations which prohibit functions which are essential for cellular growth and division are necessarily lethal. Strains carrying mutations in essential genes can, however, be isolated if the lethal effect of the mutation is only exhibited under certain environmental conditions. Under the restrictive condition, such as high or low temperature, the mutant gene product is inactive and the cells cannot grow, whereas under the permissive condition, such as optimal temperature, the protein has its wild-type activity. Thus, the mutants can be maintained at the optimal temperature and the effect of the mutations assessed at the restrictive temperature. Mutants for which temperature is the restrictive condition may be of two types. If the restrictive temperature is above that optimal for growth they are known as temperature-sensitive or *ts* mutants, whereas, if it is below the optimum, the mutants are referred to as cold-sensitive or *cs*. The isolation of conditional lethal mutants would appear to be a particularly useful strategy for studying the mitochondrial genetics of obligate aerobes. However, this has not been seriously pursued, although a cold-sensitive mutant of *Aspergillus nidulans* has been shown to be of mitochondrial origin.

A systematic search for mitochondrial mutations with a temperature-sensitive phenotype has been made in the facultative anaerobe, *S. cerevisiae*. The *ts* mutants which have been isolated can be divided into those which generate cytoplasmic petite mutants at the restrictive temperature and those which do not. The former class probably has defects which affect mitDNA replication, although it should be remembered that petite mitDNA is replicated in cells which do not have a mitochondrial protein synthesis system. The second class includes the *syn*⁻ mutants which have specific defects in the mitochondrial translation system. They were recognized during the screening of *mit*⁻ mutants as cells which did not retain mitochondrial protein synthesis but which could be distinguished from cytoplasmic petites. The distinction was made on the basis of two criteria: the ability to revert to the grande state and the formation of grande recombinants when crossed with ρ^- petites. The *syn*⁻ mutants have alterations in the genes for mitochondrial rRNAs and tRNAs.

It is necessary to isolate mutants which are temperature-sensitive since prolonged growth in the absence of mitochondrial protein synthesis itself results in the induction of petites. This was shown by Williamson and his colleagues in their studies with the mitochondrial protein synthesis inhibitor erythromycin. The fact that mitochondrial protein synthesis is necessary for the maintenance of wild-type mitDNA demonstrates its role in the accurate replication of the mitochondrial genome.

Cold-sensitive mutants have rarely been studied in any detail. However, Linnane and his co-workers have isolated a mitochondrial cold-sensitive mutant which has a defect in the assembly of the membrane-bound ATPase at the restrictive temperature. The mutation is allelic with others which determine oligomycin resistance; they all map in a gene designated *oli* 1.

Electrophoretic variants Mutations do not always result in an observable change in protein function, but these neutral or cryptic mutations can often be detected as a change in the electrophoretic mobility of a protein due to alterations in either its net charge or overall size. Butow, Perlman and their colleagues have looked for electrophoretic variants of mitochondrially-synthesized proteins in yeast. These proteins were labelled with ^{35}S in the presence of cycloheximide and the labelled products from a wide variety of strains analysed by SDS-polyacrylamide gel electrophoresis. Although no variants were detected in any proteins whose function was already known, a number of mutant forms of three other proteins were found. The three new genes thus defined have been mapped and assigned the names *var*-1-3. The *var*-1 protein is now known to be associated with the mitochondrial ribosome, although it is not essential for protein synthesis. A similar protein in the *N. crassa* mitochondrial ribosome is, however, required for translation to occur. The role of the *var*-2 protein is still unknown, but *var*-3 has been identified as the DCCD-binding lipoprotein of the inner mitochondrial membrane. Again this contrasts with *Neurospora* in which this protein is encoded by a nuclear rather than a mitochondrial gene.

Mapping the mitochondrial genome

Once several mitochondrial mutants of different kinds had been collected in one species their relative positions on the mitochondrial genome could be mapped. A wide range of techniques has been used for this purpose and a number of peculiar problems arose for which some ingenious solutions have been found.

Physical techniques The technique of restriction endonuclease mapping has been applied to a number of mitochondrial DNA molecules. Restriction endonucleases are bacterial enzymes which cleave double-stranded DNA at specific sites determined by its base sequence. Different restriction endonucleases have different sequence specificities and therefore two or more such enzymes used both separately and in combination permit a fragment map of any DNA molecule to be constructed. This method of physical mapping can be used quite independently of recombination analysis and, indeed, has been used to map the mitochondrial DNA of *Neurospora crassa*, an organism in which mitochondrial genetic recombination has not been observed.

Stable mitochondrial RNA, rRNA and tRNA molecules, may be hybridized to denatured fragments of mitDNA obtained from a restriction enzyme digest. The genes encoding these RNA molecules can, in this way, be located on the physical map of mitochondrial DNA. This method has shown that the gene for the 22S rRNA of *N. crassa* is split and the spacer between the two halves of the gene contains sequences complementary to tRNAs. The genes for 22S and 15S rRNA molecules lie adjacent to one another in *Schizosaccharomyces* mitDNA, an arrangement also found in prokaryotes and the nuclei of eukaryotes. However, in both yeast and *Tetrahymena* these two genes are far apart on the mitochondrial map and this raises the question of how they are coordinately transcribed.

Petite deletion mapping Deletion mapping, where the loss of a segment of DNA is correlated with the loss of a gene or genes, is a rapid method of determining gene order and has been used extensively in bacteriophage genetics. It has also been employed in the mapping of the mitochondrial genome of *S. cerevisiae* where the method has the novelty of using cytoplasmic petite mutants as a 'natural' source of deletions. The ρ^- petite mutants retain varying amounts of their mitochondrial DNA; some sequences are deleted whilst others are reiterated. Once a number of mutants in mitochondrial genes were available it was possible to determine, by crosses between petite and grande strains, which markers the petite DNAs had retained. An approximate genetic map could then be constructed from the pattern of co-retention and co-loss of mitochondrial genes in a wide range of petite mutants. The pattern of retention of genes coding for rRNA or tRNA could be followed by performing RNA-DNA hybridization experiments between petite DNAs and mitochondrial RNAs. These techniques defined the relative positions of RNA genes and structural genes on the mitochondrial map.

These mapping methods are rapid but not very precise; the technique was refined as knowledge of the molecular and genetic constitution of petite mutants

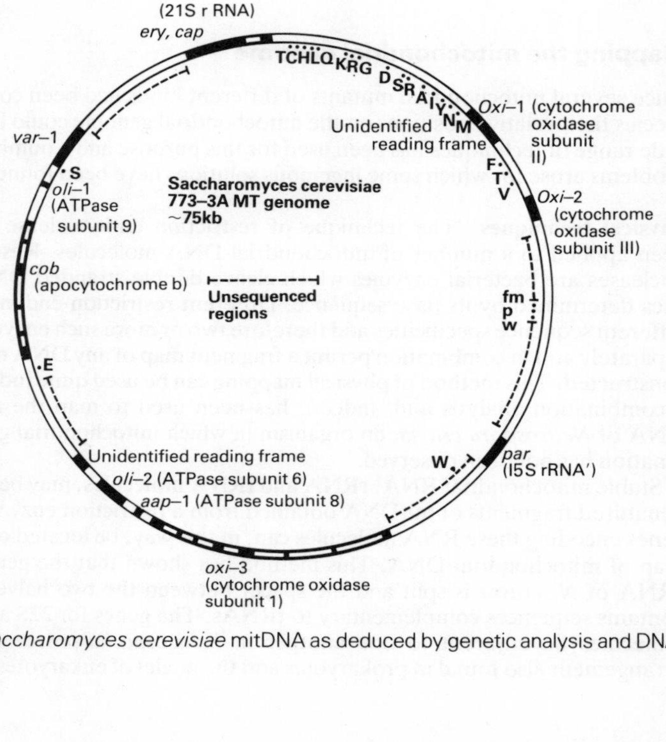

Fig. 12 The *Saccharomyces cerevisiae* mitDNA as deduced by genetic analysis and DNA sequencing.

increased. An advantage of studying petite mitDNA is that it can be isolated intact enabling unambiguous restriction maps to be constructed. The ideal petite DNAs for a more detailed genetic analysis are those which have a stably-inherited sequence with few, if any, secondary rearrangements or deletions. It is most important that this sequence should represent a single continuous segment of grande DNA. The mutants used for such studies should therefore contain only consecutive combinations of genetic markers. They are usually obtained as spontaneous petites and are then rigorously purified by repeated subcloning. The physical characterization of these petite DNAs was carried out by both restriction endonuclease analysis and DNA-DNA hybridization with the parent grande mitochondrial DNA. From these data the degree of sequence overlap between the different petite DNAs was calculated. This was correlated with the retention of genetic markers and the position of the genes on the physical map was thus determined quite accurately. Studies using variations of this basic technique have been performed in the laboratories of Borst, Linnane and Slonimski and have produced a detailed and consistent map of the yeast mitochondrial genome which is illustrated in Figure 12.

Zygotic gene rescue Mitochondrial genes which are retained by petite mutants cannot be expressed in such cells because of the absence of the organelle protein synthesizing system. These genes may, however, be retrieved in grande zygotes produced from a petite × grande cross. This technique of zygotic gene rescue permits the detection of the mitochondrial genes carried by the petite partner in such a cross without the requirement for genetic recombination. This depends on the ability of the transcription and translation system of the grande partner in the cross to bring about the expression of the genes carried by the petite partner. The investigator must, of course, have some biochemical method of identifying the protein products synthesized by the zygote cells. It was this technique that was used to map the *var*-1 gene.

Mitochondrial recombination

Classical recombination analysis has proved useless for mapping all but the most closely linked markers on the mitochondrial genome. This is due to the exceptionally high level of recombination which occurs between mitDNA molecules. In *Saccharomyces*, a separation of 1000 base pairs between two mitochondrial genetic markers means that they will recombine with a recombination frequency equivalent to that of two completely unlinked genes. A distance of 1000 base pairs, or 0.33 μm of DNA, is only 1.3% of the total length of yeast mitDNA. Therefore two mitochondrial markers must be very closely linked indeed for conventional recombination frequency analysis to be of any use at all in determining their relative position. In spite of its limited use for mapping purposes the study of mitochondrial gene recombination has provided much information about the way cells organize and maintain their cytoplasmic genomes.

Mitochondrial recombination has so far been demonstrated in *S. cerevisiae, Schizosaccharomyces pombe* and *Aspergillus nidulans*. It has not been detected, despite intensive efforts, in either *Neurospora crassa* or *Paramecium aurelia*. The most extensive study of mitochondrial recombination has been made with *S. cerevisiae* and so a detailed consideration of the phenomenon will be confined to this organism.

In Chapter 1 it was noted that the cytoplasm could contain a mixture of mitDNA molecules of different genetic constitution, a state which was described as a heteroplasmon. At the phenotypic level a mitochondrial heteroplasmon will be indistinguishable from a mitochondrial recombinant. Thus, it is essential to provide some physical evidence that mitochondrial recombination has, in fact, taken place.

This has been demonstrated in yeast for both petite × petite and petite × grande crosses. This was possible because of the buoyant density different between different petite mitDNAs and between petite and grande mitDNAs. A cross between two strains with DNAs of different buoyant densities enabled investigators to demonstrate recombination by retrieving mitDNA of unique and intermediate buoyant density from the diploid progeny of the cross. Restriction enzyme analysis has been used to demonstrate mitochondrial recombination in a cross between two grande strains. The fragment pattern produced by the cleavage of the mitDNA of the diploid progeny from the cross differed from that of either parent.

Two basic experimental protocols have been used to investigate yeast mitochondrial recombination at the genetic level. In zygote clone analysis, a micromanipulator is used to isolate successive daughter cells produced by the budding of an individual zygote cell. The other type of experiment is easier to perform but is more difficult to interpret. This is random diploid analysis in which mass matings are performed and the diploid progeny are analysed many generations later.

Electron microscopic studies of the formation of a zygote cell have revealed that mitochondrial membranes degenerate following cell fusion and that the normal structure of the mitochondria is not reformed until the zygote starts to bud. Thus, there do not appear to be any physical barriers to genetic exchange between mitochondrial DNA molecules within the zygote. In any case it seems that the mitochondrial complement of a yeast cell is a fluid continuum in which individual mitochondria frequently fuse and separate. Zygote clone analysis has shown that most mitochondrial recombination occurs in the zygote cell itself although some further recombination events take place in the early diploid buds. It has been claimed that cytoplasmic mixing in the zygote is incomplete and that buds arising from the central region, where cell fusion occurred, are more likely to be recombinant for their mitochondrial genes than those which are produced from the poles. Less cytoplasmic mixing is supposed to have occurred in the latter region and so buds of parental type are more likely to be produced. This claim has been disputed and it may be that there are strain differences in the degree of cytoplasmic mixing.

It is interesting that although a zygote cell contains about 100 mitDNA molecules, buds with a pure mitochondrial genotype (homoplasmic buds) are very rapidly produced. When crosses involving a number of markers on the mitochondrial chromosome were performed it was found that if one recombinant type was present in the progeny of the zygote cell the reciprocal recombinant type was not. Thus, mitochondrial genetic exchange appeared to operate by a process of non-reciprocal recombination or gene conversion. Another peculiarity of these recombination experiments was that in certain crosses one or the other parental genotype or a particular recombinant type appeared to be preferentially inherited. These features have been studied in more detail by random diploid analysis.

The usual protocol for this type of experiment was developed by Slonimski and his colleagues and is often referred to as the 'standard cross'. Approximately equal numbers of stationary phase cells of two strains of opposite mating type are mixed together on an agar plate. This plate contains a minimal medium which is selective for nuclear diploids but does not select for either mitochondrial genotype. More than a thousand mating events occur on the plate and growth of the resulting diploid clones is allowed to proceed until colonies are confluent. This means that the zygotes and their diploid progeny must have undergone about 20 divisions on the plate. The cells are then harvested and analysed for their mitochondrial phenotype.

The pattern of inheritance and recombination of mitochondrial genes in such a random cross showed a number of peculiarities which are described by the following special terms. *Bias* is a term applied when the frequency of recovery of the two parental genotypes or of two reciprocal recombinant types is not equal. *Asymmetry* is used when this inequality is the result of the action of a nuclear gene whereas *polarity* is applied when inequalities result from the action of a mitochondrial gene. A gene which is preferentially recovered in the progeny of a random cross due to any of these three phenomena is said to show a high degree of *transmission*.

The phenomenon of polarity has been studied in great detail and some information is now available as to its molecular basis. The first mitochondrial locus which determined the polarity of transmission of mitochondrial genes was defined by Dujon, Slonimski and their collaborators and is called omega (ω). There are three alleles of the ω locus, ω^+, ω^- and ω^n. In an $\omega^+ \times \omega^+$ or an $\omega^- \times \omega^-$ cross, the so-called homosexual crosses, both parental types are equally represented in the diploid progeny and the frequency of reciprocal recombinant types is also approximately equal. However, in a heterosexual cross, $\omega^+ \times \omega^-$, recombination frequencies are higher and this recombination occurs at the expense of the ω^- parental genotype which is transmitted to the progeny at much lower frequencies than the ω^+ parental genotype. The increased proportion of recombinants produced by the heterosexual cross favours particular recombinant types at the expense of their reciprocal forms. Strains bearing the ω^n allele produce recombination patterns typical of homosexual crosses irrespective of whether they were mated with ω^+ or ω^- partners. Representative data for homosexual and heterosexual crosses are given in Table 6.

When the phenomenon of polarity, determined by the ω locus, was first discovered it was postulated that ω^+ mitochondria were analogous to HFr strains of *Escherichia coli* and that the recombination frequencies observed in $\omega^+ \times \omega^-$ were the result of a kind of sexual conjugation between mitochondria. It was subsequently discovered, however, that the ω^+ allele only exerted its effect over a limited region of the yeast mitochondrial genome whereas the integrated F factor in HFr strains determines the polar transmission of the entire *E. coli* chromosome.

The ω locus was mapped and found to lie within the gene for the 21S mitochondrial rRNA and to promote the polar transmission of genes such as *ery* and *cap* lying on either side of it. Strains carrying the ω^+ allele have subsequently been found to have a 1050 base pair insertion sequence within the 21S rRNA gene. Neither ω^- or ω^n strains have this insertion and since such strains, when crossed with ω^+ cells, produce different transmission patterns, this indicates that the insertion, although necessary, is not, alone, sufficient to determine polarity. This is confirmed by the observation that ω^+ strains may be derived from ω^- strains by

Table 6 Characteristics of non-polar and highly polar mitochondrial genetic crosses

Parameter	Percentage of total Cross I (homosexual)	Cross II (heterosexual)
Individual genotype		
$C^RE^RA^R$	41.4	31.3
$C^RE^SA^R$	5.3	3.9
$C^SE^RA^R$	1.6	0
$C^SE^SA^R$	8.1	0
$C^RE^RA^S$	5.1	22.4
$C^RE^SA^S$	5.4	31.5
$C^SE^RA^S$	1.6	0.3
$C^SE^SA^S$	31.4	10.6
Total for single markers (transmission frequencies)		
C^R	57	89
E^R	50	54
A^R	56	35
Recombination frequency		
C, E	14	36
C, A	20	54
E, A	20	27
Recombination polarity	Ratio	Ratio
C, E	3.3	141
C, A	1.1	—
E, A	0.5	5.8

Cross I is *rho*⁺ *ome*⁺ *cap*1-r *ery*1-r *ana*1-r *par*1-r
×
rho⁺ *ome*⁺ *cap*-s *ery*-s *oli*-s *ana*-s *par*-s
Cross II is *rho*⁺ *ome*⁺ *cap*1-r *ery*1-r *oli*1-r ana1-r *par*1-r
×
rho⁺ *ome*⁻ *cap*-s *ery*-s *oli*-s *ana*-s *par*-s
Symbols: C = *cap*-1, E = *ery*-1, A = *ana*-1, R = resistant, S = sensitive.
(Adapted from Nagley *et al.*, 1977.)

mutations which are probably deletions. Borst and Grivell have proposed that there is a second, but much smaller, insertion sequence in ω^- strains which is absent in ω^+ cells. It is suggested that ω^n strains carry only part of this insertion. The effect of these insertion sequences is to promote non-reciprocal recombination between the ω^+ and ω^- strains. This occurs by a process of asymmetric gene conversion in which the DNA sequences in the region of the ω locus in the ω^- strain are replaced by copies of the homologous sequence from the ω^+ strain. The conversion, or replacement, event is initiated at the region of non-homology which the site produces between the paired parental DNA strands. The probability of genes on the ω^- DNA being replaced by copies of genes from the ω^+ parent falls off with increasing distance, in either direction, from the ω locus.

The limitation of the effect of the ω locus to a small region of the mitochondrial genome rendered analogies with bacterial sex invalid and it was suspected that the phenomenon was unique. However, it seems that there are other regions of the yeast mitochondrial genome where polarity of gene transmission occurs and that these are also due to gene conversion events promoted by insertion sequences. A very thorough genetical study of gene conversion at the *var*-1 locus has been made by Butow, Perlman and their colleagues. The phenotype of *var*-1 mutants is determined by performing polyacrylamide gel electrophoresis on proteins labelled in the presence of cycloheximide. Crosses between strains producing slow-migrating (S) and fast-migrating (F) electrophoretic variants of the *var*-1 protein produced two types of recombinant both having proteins of intermediate electrophoretic mobility. The two types were intermediate fast, IF, and intermediate slow, IS. Reciprocal recombinants were not produced in equal proportions and this pattern of asymmetric gene conversion was explained when it was found that the S-form of the *var*-1 gene contained two discrete insertion sequences, of 36 base pairs and 57 base pairs, which were absent from the F gene. It was postulated that asymmetric gene conversion operated to add an insertion into the DNA of the partner which lacked it when the heteroduplex was formed. The scheme for the pattern of recombination in all possible pair-wise crosses is shown in Figure 13.

Experiments with yeast have revealed a puzzling aspect of mitochondrial genetics. Mitochondrial mutants can easily be isolated and homoplasmic clones rapidly segregate from zygote cells in spite of the fact that a haploid cell contains about 50, and a zygote about 100 molecules of mitDNA. Biochemical analysis thus shows the mitochondrial genome to be highly polyploid whereas genetic analysis provides a much lower figure of tetraploid to hexaploid. Several attempts have been made to explain this difference between the number of genetically active copies of the yeast mitochondrial genome and the actual number of DNA molecules. Some of these explanations have suggested that there are one, or a few, master copies of the mitDNA which are the only ones able to replicate. The other 'slave' copies are able to express but not replicate their genes. Existing biochemical evidence indicates that all the mitDNA molecules are able to replicate and current theories stress the ability of non-reciprocal recombination to spread mutations and enforce genetic uniformity on the multiple mitochondrial genomes. It is thought that there are no barriers to the recombination of any one mitDNA molecule in a vegetative cell, or in a zygote, with any other such molecule. The 50 to 100 molecules of mitDNA are said to form a panmictic or completely mixed

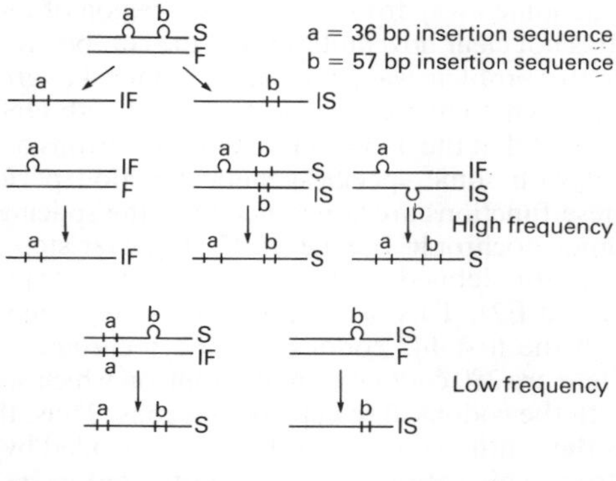

S = Slow-migrating variant } Parental types
F = Fast-migrating variant
IS = Intermediate slow-migrating variant } Recombinant types
IF = Intermediate fast-migrating variant

Fig. 13 Gene conversion at the *var* locus.

pool which may take part in multiple rounds of genetic recombination. This idea is supported by the finding, from the electron microscopic examination of serial thin sections, that vegetative yeast cells contain very few mitochondria; between one and six.

Further work, using a fluorescent probe to reveal mitochondrial DNA *in situ*, has shown that the number of clusters of mitochondrial DNA (the number of mitochondrial 'nuclei' or 'chondriolites') is similar to the number of physically distinct mitochondria.

The mitochondrial complement of a single cell is now believed to be a dynamic system in which individual organelles are continually fusing and being regenerated by fission. The DNA molecules of the chondriolites are constantly recombining both with one another and with those in other chondriolites. These recombination events commonly involve gene conversion and genetic homogeneity between the multiple copies may rapidly be achieved. Current studies are attempting to define the rules which govern the direction of these conversion events and the work with both the ω and *var*-1 loci suggests that insertion sequences play a major role.

A number of other mitochondrial genes in yeast have a very complex structure. The most extraordinary is the so-called *cob-box* region of the genome. Mutations in this region are often pleiotropic and produce defects in the synthesis of both cytochrome b and cytochrome c oxidase. The region appears to determine the synthesis of both the cytochrome b apoprotein and subunit 1 of the cytochrome c oxidase complex. The cob-box region comprises 11 genetic loci. Seven of these loci (E1–E7) are all 'exons' and actually encode different parts of the cytochrome b aproprotein. The other six loci (I1–I6) are regarded as intervening sequences or 'introns', which do not encode any part of the protein molecule. The primary transcript from the gene is an mRNA precursor which contains both intron and exon sequences. The intron sequences are removed and the exons joined together by a process known as 'splicing' in order to create the mature mRNA which is translated into protein. It is mutations in the six introns that are pleiotropic, affecting the synthesis of both cytochrome b and cytochrome c oxidase. The structural gene for subunit 1 of the cytochrome c oxidase complex, *oxi*-3, is about a quarter of the genome away from the cob-box region of the yeast mitochondrial genetic map. It is not clear how mutations in the cob-box region can affect it.

A solution to this problem was provided by Slonimski's group in a brilliant series of experiments which combined classical genetics with DNA sequence analysis. Slonimski proposed that the mutations within the introns of the apocytochrome b gene interfered with some specific genetic function performed by the intron sequences. These functions are concerned with the splicing of the primary transcript of the apocytochrome b gene itself. The first step in this process is the removal of the intron defined by the I1 locus and results in the fusion of the first two exons (E1 and E2). This fusion produces a long open reading frame which extends through the first 143 codons of apocytochrome b and into the second intron (I2). There are 280 codons within this intron which specify amino acids and are in phase with the codons in the upstream exons. Thus, the excision of the first intron permits the synthesis of a hybrid protein encoded by the E1 and E2 exons and the I2 intron. This hybrid protein is postulated to be a 'maturase' enzyme which is involved in the excision of the I2 intron. This maturase is responsible, therefore, for the destruction of its own mRNA and is thus self-regulating. Although mutations within the I2 maturase gene are pleiotropic, the effect appears to be indirect. Further processing of the transcript (which is blocked in I2 mutants) leads to expression of a second maturase, coded by I4, and it is this maturase that has the dual role—splicing of both its own intron and of introns present in *oxi*-3. Figure 14 details these interactions.

The accumulation of sequence data for a large number of mitochondrial genes has demonstrated that the maturase hypothesis is applicable to the apocytochrome b and cytochrome oxidase subunit 1 genes of other fungi. The theory is not universally applicable and many mitochondrial introns do not contain protein encoding sequences. The nucleotide sequences of these introns have been examined to assess their potential for intra-strand base pairing. These studies have demonstrated that many mitochondrial introns display a striking conservation of secondary structure. Davies and his co-workers have proposed a model whereby RNA base-pairing within these introns forms a stable 'core' structure which brings the two ends of each intron close together, permitting additional base-pairing to occur, and results in the precise alignment of the two splice points (Figure 15). A number of nuclear gene introns can adopt a similar secondary structure, including the intron within *Tetrahymena* RNA which can be spliced *in vitro* in the absence of any proteins. The torsional stresses improved by the base-pairing within the intron apparently give the RNA enzyme-like properties which facilitate the breakage and rejoining of the polynucleotide chain.

Several yeast mitochondrial genes have now been sequenced, including *oxi*-1, -2 and -3 (cytochrome c oxidase subunits 2, 3 and 1), cob-box (apocytochrome b), *oli*-1 and -2 (ATPase subunits 9 and 6), *aap*-1 (ATPase subunit 8) and *var*-1 (a ribosomal protein). In all cases comparison of the sequence of the gene with the amino acid sequence of the protein has revealed that yeast mitochondria employ a different genetic code to that used by both procaryotes and eucaryotes. For

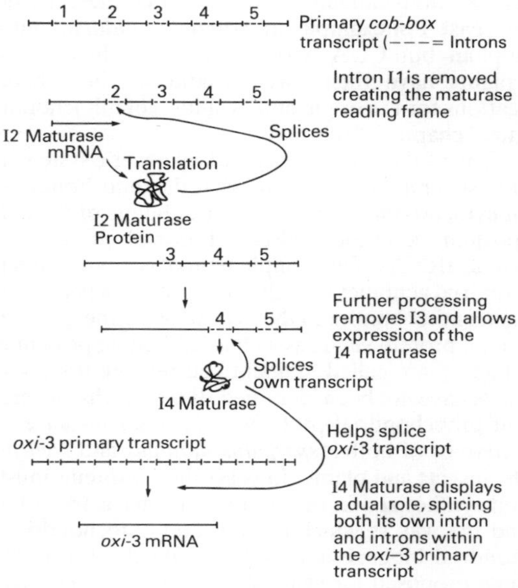

Fig. 14 The maturase proteins and the processing of *cob-box* and *oxi*-3 mRNAs.

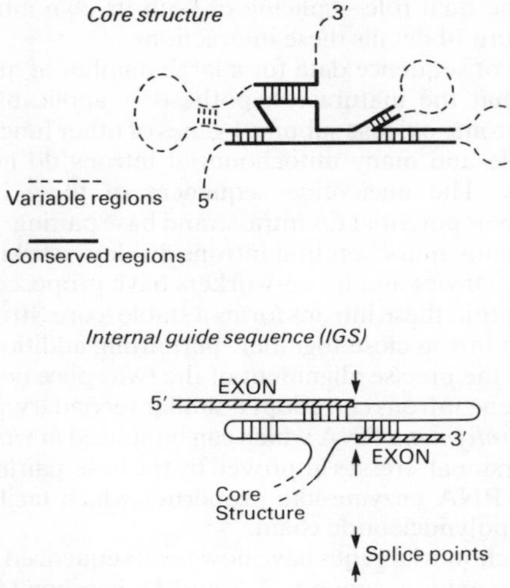

Fig. 15 Intron splicing model.

example CUA, which usually specifies leucine, codes for threonine in yeast mitochondria. In addition AUA, usually isoleucine, codes for methionine, and UGA, a termination codon in the universal system, specifies tryptophan. Other fungal mitochondria also use non-standard codes, but often these are distinct from one another and from yeast. For example, in *Aspergillus nidulans* mitochondria UGA codes for tryptophan but CUA and AUA retain their normal specificities; mammals and plants have their own variations. These findings have very important implications for the origin and evolution of mitochondria which will be discussed in the final chapter of this book.

The intensive study of the yeast mitochondrial genetic system has yielded what is probably an almost complete picture of what the mitochondrial DNA encodes: three subunits of cytochrome c oxidase, three subunits of the ATPase, the cytochrome b apoprotein, a mitochondrial ribosomal protein, 2 mitochondrial rRNAs, and about 25 tRNAs. This complement of genes may not in fact be typical of fungal mitochondrial genomes as additional genes are present in other systems. *Aspergillus nidulans* mitochondrial DNA contains all the genes present in yeast, except the ribosomal protein *var*-1, as well as additional protein-encoding genes. These additional genes are called ‘unidentified reading frames’ (URFs) as their translation products have not been identified. It is clear that several of these URFs encode important mitochondrial proteins: URF1 for instance is present in the mitochondrial genomes of both *Aspergillus nidulans* and *Neurospora crassa*, as well as mammals, insects and plants. In yeast these proteins must be encoded by nuclear genes and synthesized in the cytoplasm. This is true of many mitochondrial proteins and the way in which nucleus and mitochondrion interact to coordinate mitochondrial biogenesis is still very much of a mystery. Regulatory schemes have been proposed for yeast, *Neurospora* and *Tetrahymena*, but such evidence as exists for them is indirect. The next phase of intense research activity in the field of mitochondrial genetics can be expected to concentrate on this problem of nucleo-mitochondrial interactions.

Summary

The construction of a functional mitochondrion requires the products of both the nuclear and the mitochondrial genomes. Although mitochondrial DNA encodes only a few proteins, these are essential for the correct integration into the mitochondrial membrane of some respiratory enzymes, encoded in the nucleus and synthesized in the cytoplasm. This role of mitochondrially-synthesized proteins accounts for the pleiotropic effect of mitochondrial mutations such as 'poky' in *Neurospora* and 'petite' in *Saccharomyces.* The mitochondrion is shown not only to have its own distinct DNA genome but also mechanisms of transcription and translation which are quite separate from those which express nuclear genes. These mechanisms show many similarities to those found in bacteria but also in other respects differ from those of both bacteria and the eukaryotic cytoplasm.

References

BORST, P. and GRIVELL, L. A. (1978). The mitochondrial genome of yeast. *Cell* 15: 705–23.

DAVIES, R. W., WARING, R. B., RAY, J. A., BROWN, T. A. and SCAZZOCCHIO, C. (1982). Making ends meet: a model for RNA splicing in fungal mitochondria. *Nature* 300: 719–24.

EPHRUSSI, R. (1953). Nucleo–Cytoplasmic Relations in Microorganisms. Clarendon Press, Oxford.

GILLHAM, N. W. (1978). Organelle Heredity. Raven Press, New York.

GOLDBACH, R. W., BOLLEN-DE BOER, J. E., VAN BRUGGEN, E. F. J. and BORST, P. (1979). Replication of the linear mitochondrial DNA of *Tetrahymena pyriformis. Biochim. Biophys. Acta* 562: 400–17.

GRIVELL, L. A. (1983). Mitochondrial DNA. *Scientific American* 248: 60–73.

LAZOWSKA, J., JACQ, C. and SLONIMSKI, P. P. (1980). Sequence of introns and flanking exons in wild-type and *box*3 mutants of cytochrome b reveals an interlaced splicing protein coded by an intron. *Cell* 22: 333–48.

MACINO, G., SCAZZOCCHIO, C., WARING, R. B., BERKS, M. M. and DAVIES, R. W. (1980). Conservation and rearrangement of mitochondrial structural gene sequences. *Nature* 288: 404–6.

MITCHELL, M. B. and MITCHELL, H. K., (1952). A case of 'maternal' inheritance in *Neurospora crassa. Proc. Natl. Acad. Sci. U.S.A.* 38: 442–9.

NAGLEY, P., SRIPRAKASH, K. S. and LINNANE, A. W. (1977). Structure, synthesis and genetics of yeast mitochondrial DNA. *Adv. Microb. Physiol.* 16: 157–277.

OLIVER, S. G. (1977). On the mutability of the yeast mitochondrial genome. *J. Theoret. Biol.* 67: 195–201.

PERLMAN, P. S., DOUGLAS, M. G., STRAUSBERG, R. L. and BUTOW, R. A. (1977). Localisation of genes for variant forms of mitochondrial proteins on mitochondrial DNA of *Saccharomyces cerevisiae. J. Mol. Biol.* 115: 675–94.

TEWARI, K. K., JAYARAMAN, J. and MAHLER, H. R. (1965). Separation and characterisation of mitochondrial DNA from yeast. *Biochem. Biophys. Res. Commun.* 21: 141–8.

THOMAS, D. Y. and WILKIE, D. (1968). Recombination of mitochondria drug resistance factors in *Saccharomyces cerevisiae. Biochem. Biophys. Res. Commun.* 30: 368–72.

THOMAS, D. Y. and WILLIAMSON, D. H. (1971). Products of mitochondrial protein synthesis in yeast. *Nature New Biology* 233: 196–9.

TZAGOLOFF, A., MACINO, G. and SEBALD, W. (1979). Mitochondrial genes and translation products. *Ann. Rev. Biochem.* 48: 419–41.

WRIGHT, R. E. and LEDERBERG, J. (1957). Extranuclear transmission of yeast heterokaryons. *Proc. Natl. Acad. Sci. U.S.A.* 43: 919–23.

3 Chloroplast genetics

Classical genetic analysis of the chloroplast system began in the early 1950s using the green alga *Chlamydomonas reinhardtii*. These studies were stimulated by the discovery that chloroplast genetic markers could be distinguished by their unique mode of inheritance, and by experimental advances that allowed their segregation and recombination to be studied. In photosynthetic organisms the bulk of the extranuclear markers map in the chloroplast genomes because these genomes generally contain more genes than their mitochondrial counterparts. This relative complexity makes the chloroplast genome a challenging system to study and only during the last few years has any sort of correlation between the genetic and biochemical data become possible. Two schools of thought have arisen concerning the molecular explanations for events analysed genetically and disagreement exists over several aspects of the subject.

Inheritance patterns for chloroplast genes

The first step in the study of *Chlamydomonas* extranuclear genetics was the isolation in 1954 by Sager and her colleagues of two streptomycin resistant strains that displayed different patterns of inheritance. The first mutation, *sr*-1, conferred resistance to low levels of streptomycin and displayed typical Mendelian inheritance. The second mutation, *sr*-2, was characterized by resistance to greater concentrations of the antibiotic and was uniparentally inherited. When an *sr*-2 mt^+ strain was crossed with a streptomycin sensitive mt^- stock, all of the haploid cells produced by meiosis in the zygote displayed resistance. Conversely, if the *sr*-2 marker was carried by the mt^- parent then all the haploid daughters were streptomycin sensitive. In backcrosses of the F_1 and succeeding progeny the same uniparental pattern of *sr*-2 inheritance was seen (Figure 16).

Several other uniparentally inherited mutations have been isolated since this pioneering work with the *sr*-2 locus. In addition to antibiotic-resistant mutants these include streptomycin and neamine dependent strains, temperature-sensitive mutants and non-photosynthetic types that can only grow in the presence of a reduced carbon source such as acetate (Table 7).

An invariant pattern of uniparental inheritance would prevent any kind of segregational or recombinational analysis of chloroplast genes. However, the 4:0 inheritance pattern is not always found and, in any cross, a very small number of 'exceptional zygotes' is produced. Most of these display biparental transmission of extranuclear markers, so that in the cross described above (*sr*-2 mt^+ × *ss* mt^-) a small number of streptomycin-sensitive (*ss*) cells are found. Analysis of the meiotic daughter cells arising from exceptional zygotes and of their mitotic progeny showed that the daughter cells carried both parental chloroplast genomes and were therefore cytoplasmic heterozygotes or heteroplasmons. Normally segregation occurs during one of the early mitotic divisions of the daughter cells so that cytoplasmically homozygous progeny are eventually produced. Occasionally

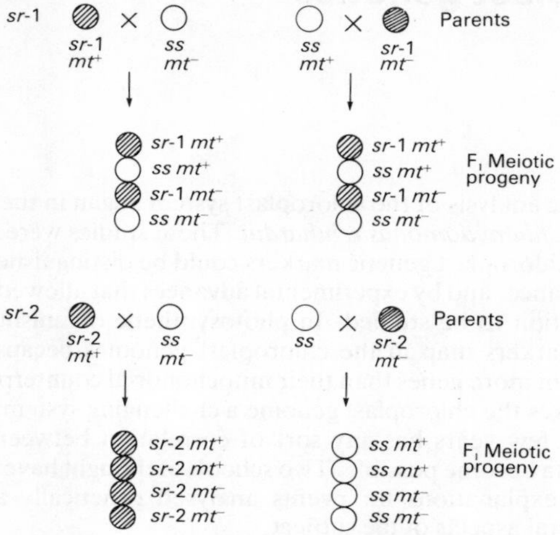

Fig. 16 Inheritance of *sr*-1 and *sr*-2 in *Chlamydomonas reinhardtii.*

a 'persistent heterozygote' will arise; a cell line which, although haploid for nuclear genes, never achieves cytoplasmic homozygosity.

In addition to these exceptional zygotes showing biparental gene transmission some 'paternal zygotes' may be produced at even lower frequencies. Such zygotes are distinguished by the fact that only markers present in the mt^- parent are found in the progeny.

The small number of exceptional zygotes normally obtained in any cross is a hindrance to the analysis of extranuclear genes but their proportion may be dramatically increased by irradiating the mt^+ parent with a sublethal dose of uv

Table 7 Some uniparental *Chlamydomonas* mutants

Gene	Mutant phenotype
ac	acetate-requiring
tm-1, *tm* 3-9	cannot grow at 35°C
tm-2	grows at 35°C only with streptomycin
ti	forms tiny colonies on all media
ery	erythromycin-resistant
kan	kanamycin-resistant
spc	spectinomycin-resistant
spi	spiramycin-resistant
ole	oleandomycin-resistant
car	carbamycin-resistant
ele	eleosine-resistant
ery	resistant to erythromycin, carbamycin, oleandomycin and spiramycin
sr (= sm)	streptomycin-resistant
sd	streptomycin-requiring

light prior to mating. Under the right conditions almost complete biparental inheritance can be achieved. This technique has allowed the unequivocal demonstration of recombination between chloroplast genes. This can be illustrated by a cross involving two cytoplasmic factors. When an acetate-requiring streptomycin-sensitive parent (*ac*$^-$ *ss*) is mated with a photosynthetic, streptomycin-resistant parent (*ac*$^+$ *sr*-2), the segregant cells from exceptional zygotes include progeny showing the parental genotypes, a few wild types (*ac*$^+$ *ss*) and a few double mutants (*ac*$^-$ *sr*-2). Assuming that there is only one type of chloroplast genome, so that the *ac*$^-$ and the *sr*-2 genes are located on the same DNA molecule, then the appearance of wild types and double mutants in such a cross must result from genetic recombination.

Analysis of somatic segregation

The discovery of recombination between cytoplasmic genes led to the expectation that a genetic map of the chloroplast genome could be deduced. However, before this could be done it was necessary to devise ways of analysing in detail the segregation of chloroplast markers during the meiotic and post meiotic divisions of exceptional zygotes. Two techniques have been developed: the octospore daughter analysis system devised by Sager and associates and the zygote clone analysis technique, methodologically identical to zygote clone analysis of yeast, perfected by Gillham and colleagues (see Chapter 2).

In octospore daughter analysis (Figure 17) germinating zygotes are transferred individually to petri dishes and one post meiotic division takes place. The eight daughter cells (octospores) are carefully separated in such a way that sister cells can be readily identified at a later stage. A further mitotic division produces sixteen cells which are isolated, grown as separate colonies and their genotypes

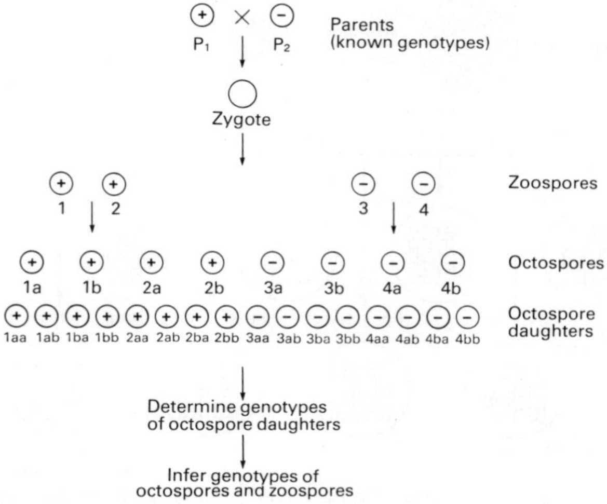

Fig. 17 Octospore daughter analysis.

analysed. If segregation of markers is detected then the octospore daughter cell was a cytoplasmic heterozygote. If no segregation is seen, then the octospore daughter must have been a pure parental type, that is, segregation occurred during the meiotic or one of the first two mitotic divisions.

The alternative system of segregation analysis, the zygote clone method, involves examination of genotypic ratios in diploid cells after numerous cell divisions. In the example shown in Figure 18 the mt^+ parent which is resistant to spectinomycin but sensitive to erythromycin and streptomycin is crossed with an mt^- strain which is spectinomycin sensitive but erythromycin and streptomycin resistant ($mt^+sp^rery^+sr^+ \times mt^-sp^+ery^rsr^r$). Biparental exceptional zygotes produce colonies resistant to all three drugs whereas cells from uniparental

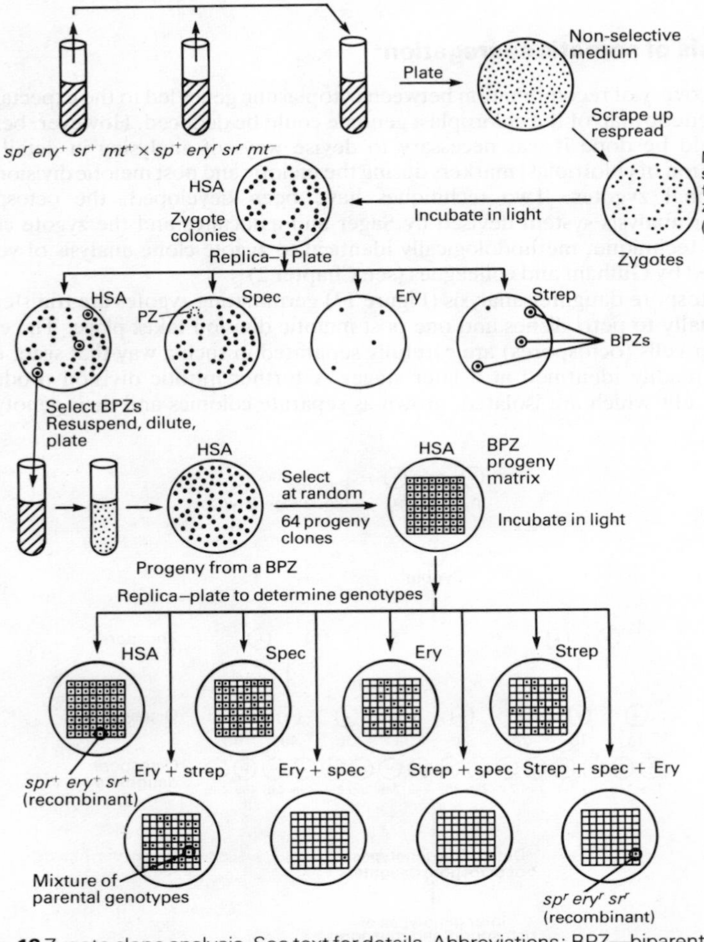

Fig. 18 Zygote clone analysis. See text for details. Abbreviations: BPZ—biparental zygote; PZ—parental zygote.

zygotes will be resistant only to spectinomycin. Colonies identified as biparental are plated out a second time. The colonies that grow are derived from the individual cells present in the original biparental zygote clones. These are now tested on selective media to work out individual genotypes and the frequency of parental and recombinant genotypes among the products of somatic segregation calculated.

Both techniques have proved to be extremely useful in allowing chloroplast gene maps to be built up, but each has disadvantages. The octospore daughter method does not allow a complete analysis of segregation in any particular cross, as segregation has not proceeded very far by the 16-cell stage. This makes it difficult to use the method to deduce recombinational frequencies. Another problem is that it is a time-consuming method in which only a few individual zygotes can be analysed. The zygote clone test, on the other hand, allows a relatively large number of zygotes to be studied and provides the genotypic ratios at the end of the segregation process. It would appear to provide a more reliable measure of recombination frequency but suffers from the problem that aberrant events may go undetected. For example, the abortion of one or more of the components of a tetrad so that less than four zoospores are produced will have a profound effect on the final genotypic ratios.

Mapping of chloroplast genes

Several different approaches to chloroplast gene mapping have been employed, mainly by Sager, and Gillham and colleagues, using data from octospore daughter and zygote clone analyses, respectively. Although the different methods have produced maps that agree in gene order, the exact spacing between different loci is not consistent between the various procedures and this has led to a certain amount of controversy over which is the 'best' method to use.

Octospore daughter analysis is analogous to the use of meiotic tetrad analysis for the mapping of chromosomal genes, especially in fungi. These two methods share the problem that recombination events occur which cannot be detected. There are two further complications in the case of chloroplast genes. A large proportion of the octospore daughters derived from exceptional zygotes will be heteroplasmons and it is difficult to know how to treat such cells in the calculation of recombination frequencies. A second difficulty is that this analysis assumes that the chloroplast of *Chlamydomonas* is genetically diploid. There is no evidence for this; the chloroplast is known to contain about 100 DNA molecules.

Sager has employed three different methods to construct chloroplast genetic maps from octospore daughter data. In the first, the individual genes are mapped according to the frequency with which recombination occurs between them and a hypothetical 'attachment point'. This point is functionally equivalent to a centromere and is conceived as the place where the chloroplast DNA is attached to one of the internal organellar membranes. This method further assumes that each daughter cell receives a chloroplast genome derived from each of the parents in the cross. In the second octospore daughter method, the distance between pairs of genes is determined from recombination frequency in the classical manner. This technique involves fewer assumptions than the attachment point method, but the problem of how to treat the heteroplasmons remains.

Cosegregation analysis is the final technique involving octospore daughter cells.

It involves the least assumptions and is the most easily justified. The degree of linkage between two chloroplast markers is estimated from the frequency with which they segregate together at the same cell division. The more closely two markers are linked, the more likely they are to cosegregate. This method should yield a consistent map. The relative positions of genes deduced by considering the cosegregation of pairs of markers can be reinforced by extending the analysis to look at the cosegregation of runs of three, four or even five markers. This method allows one to assign map positions with some confidence but map distances are less certain due to the possibility of multiple rounds of recombination.

In the alternative approach to mapping chloroplast genes, zygote clone analysis, multiple rounds of recombination are inevitable. Nevertheless, this technique represents the most powerful and reliable method available because the large number of divisions involved in the production of a zygote clone produce a population of cells in which the proportion of heteroplasmons is virtually zero. This allows reliable recombination frequencies to be calculated. The method has been used to produce a linkage map for several extranuclear markers which compares well with the consensus map derived from the three octospore daughter methods (Figure 19).

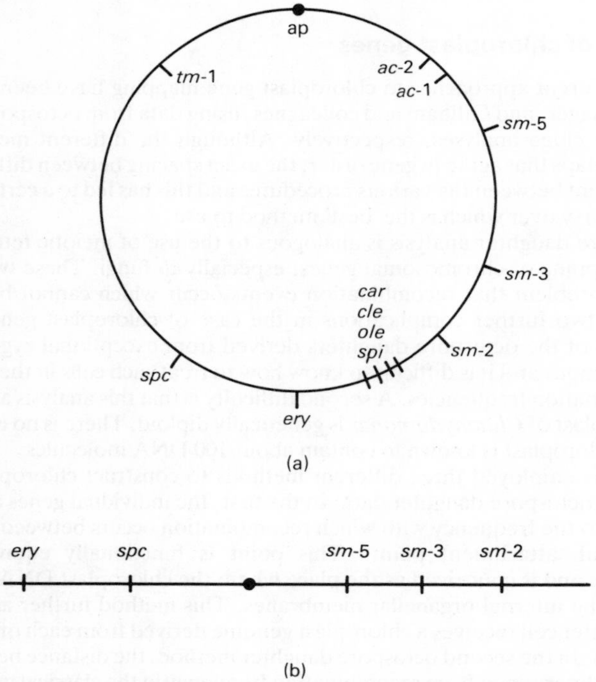

Fig. 19 Genetic maps of the *Chlamydomonas reinhardtii* chloroplast genome. (a) Circular map derived by octospore daughter analysis. (b) Linear map from zygote clone analysis. Abbreviations: ap—attachment point; gene designations—see Table 7.

Molecular biology of chloroplasts

At the same time that linkage maps for chloroplast genes were being constructed other researchers were employing biochemical techniques to study the chloroplast genetic system. The result is that, as with mitochondria, a substantial amount is now known about the molecular nature of the chloroplast genome and how it is transcribed and translated.

Chloroplast DNA The first evidence that chloroplasts might contain DNA was provided by Chiba in 1951, who showed that organelles in the clubmoss *Selaginella* and in the leaves of two flowering plants contain material that gives a positive reaction with the DNA-specific Feulgen stain. Similar experiments, performed with *Chlamydomonas,* demonstrated several bodies within the chloroplast that were stained by both the Feulgen reaction and acridine orange. These staining reactions could be abolished by pretreatment of the cells with DNase. Electron microscopy subsequently revealed DNase-sensitive fibrils within the chloroplast of *Chlamydomonas*.

These cytological demonstrations of DNA molecules in chloroplasts were followed by attempts to isolate this DNA in an intact form. It was found that chloroplast DNA from *Chlamydomonas* could be prepared as a satellite band on caesium chloride gradients of whole cell DNA, as its buoyant density (1.695 g cm^{-3}) is significantly less than that of nuclear DNA (1.721 g cm^{-3}). This difference is due to the low GC content of *Chlamydomonas* chloroplast DNA (35.7%) compared with that of its nuclear DNA (64.3%). Most algal chloroplast DNAs have a relatively low proportion of GC, allowing them to be purified from nuclear DNA by density gradient centrifugation, a technique that is not always successful with higher plants in which the GC contents of chloroplast and nuclear DNAs are often similar.

Chloroplast genomes of algae and higher plants often differ in size. The land plants that have been studied so far have a chloroplast DNA of about 150 kb whereas the genome size in algae is highly variable. It ranges from 85 kb in *Codium fragile* to possibly 2000 kb in *Acetabularia*, with the chloroplast DNAs of *Euglena gracilis* and *Chlamydomonas reinhardtii* being similar to those of higher plants at 150 and 190 kb, respectively. A comparison of some of the properties of algal chloroplast DNAs is given in Table 8.

All chloroplast DNAs that have been examined are circular; unlike mitochondrial DNA there are no linear genomes. The number of molecules per organelle is difficult to determine but seems to be in the order of 70 to 100 for most algae,

Table 8 Some properties of algal and plant chloroplast DNAs

Organism	Genome size μm	Kb	% GC content chloroplast genome	nuclear genome
Euglena gracilis	40	150	25	48
Chlamydomonas reinhardtii	62	190	36	64
Chlorella ellipsoidea	—	175	36	56
Acetabularia	200	2000	—	—
Lettuce	43	150	38	38

including *Chlamydomonas* and *Euglena*. An exception is *Acetabularia* which possesses only two copies of the 2000 kb molecule per organelle. In *Chlamydomonas* and *Euglena* the chloroplast DNA seems to be replicated at a different stage of the cell cycle to the nuclear DNA. In density labelling experiments using cultures synchronized on a 12 hour light–12 hour dark cycle each cell divided mitotically near the end of the dark period. Incorporation of ^{15}N-nucleotides into nuclear DNA occurs during the second half of the dark period, indicating that the chromosomes replicate shortly before cell division. Chloroplast DNA, on the other hand, appears to replicate during the early and middle parts of the light period, a few hours after cell division. However, artefactual results are not uncommon with synchronized cultures so these data should be treated with some caution.

The replication process for a single chloroplast DNA molecule has not been completely elucidated in any organism. A system similar to the D-loop mechanism of mitochondrial DNA replication (see Figure 11a) has been proposed for the chloroplast DNA of *Euglena* on the basis of electron microscopic observations and a single replication origin has been located on this molecule. A similar mode of replication has also been suggested for chloroplast DNAs of other organisms, although larger molecules may have more than one replication origin and the mechanism may not be straightforward as in some organisms there is evidence for both D-loop and rolling-circle modes of replication.

Chloroplast rRNA molecules Chloroplast ribosomes of higher plants usually contain four rRNA components: a single molecule of about 16S in the small subunit and three rRNAs of 23S, 5S and 4.5S in the large subunit. This is the typical prokaryotic arrangement except for the presence of the 4.5S rRNA, which has no equivalent in bacteria. This molecule is absent from the chloroplast ribosomes of *Euglena gracilis* which therefore has exactly the same rRNA complement as bacteria. *Chlamydomonas* chloroplasts have a unique arrangement of rRNAs; the small ribosomal subunit contains the usual 16S rRNA but a total of four different molecules are found in the large subunit—23S, 5S, 7S and 3S (Table 9).

Each of the chloroplast rRNA molecules hybridizes to the chloroplast DNA of the same organism showing that genes for these molecules are located on the organelle genome. The synthesis of chloroplast rRNAs is very similar to that of *E. coli*, with the rRNA genes clustered together and probably transcribed to give a single precursor molecule that is subsequently processed to yield the mature rRNA species. Such precursor molecules have been detected in *Euglena*.

Table 9 Sedimentation coefficients (S-values) for *E. coli* and chloroplast rRNA molecules

Organism	rRNAsizes Large subunit	 Small subunit
E. coli	23S, 5S	16S
Chloroplasts		
Higher Plants	23S, 5S, 4.5S	16S
Euglena gracilis	23S, 5S	16
Chlamydomonas reinhardtii	23S, 7S, 5S, 3S	16S

Chloroplast tRNA molecules Fractionation of chloroplast tRNAs by two-dimensional polyacrylamide gel electrophoresis resolves about 35 spots for *Euglena*, sufficient tRNAs to decode the entire genome using normal codon-anticodon pairing rules. Hybridization of tRNA preparations to the chloroplast genome of *Euglena* indicates nine different loci where tRNA genes may be found, several of which have now been sequenced. However, it is still not known whether these nine loci contain genes for all the tRNA molecules in the chloroplast genetic system, although it is considered unlikely that import from the cytoplasm occurs.

Those chloroplast tRNA molecules that have been sequenced are very similar to their prokaryotic counterparts, displaying most of the invariant bases seen in bacterial molecules. All chloroplast genetic systems seem to use the universal genetic code without any of the variations noted for mitochondria.

Chloroplast mRNA molecules Very little is known about chloroplast mRNAs beyond the fact that some are polyadenylated whereas others are not. It had been suggested that mRNAs for soluble polypeptides are translated on free ribosomes in the stroma, whereas polypeptides that reside within the thylakoids are synthesized on membrane-bound polysomes. This now seems unlikely; a major thylakoid protein has been found to be translated on free ribosomes.

Chloroplast RNA polymerase The *Chlamydomonas* chloroplast RNA polymerase is similar to the *E. coli* enzyme in complexity, comprising at least six subunits involving five different polypeptides of molecular weights 39,000, 51,000, 52,000, 93,000 and 96,000. None of these proteins is thought to be encoded by the chloroplast genome. The *Chlamydomonas* chloroplast enzyme also resembles that of *E. coli* in being sensitive to rifampicin, resistant to α-amanitin and in requiring an additional polypeptide, the sigma protein, for correct recognition of promoter sequences.

Chloroplast ribosomes The chloroplast ribosomes of spinach were the first organellar ribosomes to be discovered, by Lyttleton in 1962. It was immediately recognized that the sedimentation coefficient of these ribosomes was closer to the typical prokaryotic rather than eukaryotic value. The intact *Chlamydomonas* ribosome is 68 to 70S in size, with the two subunits sedimenting at 53S and 38S. These ribosomes contain the rRNA molecules described previously as well as 33 proteins in the large subunit and 31 in the small subunit, about the same number as in *E. coli* ribosomes. The *Euglena* chloroplast ribosome is about the same size but the two subunits sediment at slightly lower values, 48S and 30S.

The similarity between the physical structure of the bacterial and chloroplast ribosomes is borne out by their ability to form hybrids, e.g. between the *Euglena* chloroplast small subunit and the *E. coli* large subunit. Such a hybrid will efficiently translate an artificial poly-U message into polyphenylalanine using *E. coli* soluble factors, whereas hybrids between prokaryotic and cytoplasmic eukaryotic subunits are inactive. Further similarities are revealed when antibiotic resistant mutants are examined. *Chlamydominas* mutants resistant to spectinomycin, streptomycin and neamine have an altered small subunit in their chloroplast ribosomes whereas resistance to erythromycin, carbomycin and cleocin involves the large subunit. Equivalent bacterial mutants have similar alterations in their ribosomal subunits. A detailed analysis of these uniparental *Chlamydomonas* mutants which affect the chloroplast ribosome has indicated the nature of the chloroplast gene product. Isolation of chloroplast ribosomes from a spectinomycin resistant strain revealed that these sediment at 66S, rather lower than the normal value. When these ribosomes were denatured and the protein complement fractionated on a polyacrylamide gel, one band was missing. In this way the mutation carried by this strain was tentatively mapped to a protein normally present in the small subunit. The streptomycin resistant mutant *sr*-2 has an alteration in a different small subunit protein whereas erythromycin resistance often results from altered large subunit proteins. While these examples demonstrate that a number of proteins of the chloroplast ribosome are encoded by chloroplast DNA many of the ribosomal proteins are coded, not by chloroplast genes, but by genes in the nucleus.

Chloroplast translation products The chloroplast genomes of algae such as *Euglena* and *Chlamydomonas* are over twice the size of the yeast mitochondrial genome and could therefore code for substantially more polypeptides than the mitochondrial system. The most direct way to determine how many polypeptides are synthesized in the chloroplast is to prepare subcellular fractions containing just the chloroplast genetic system. Intact chloroplasts can be isolated from several higher plants although, amongst the algae, this method has not been universally successful. Protein synthesis by isolated chloroplasts can be switched on by placing the preparation in the light, and newly synthesized polypeptides can be detected by supplying labelled amino acids as substrates. These experiments have shown that a variety of stromal proteins and polypeptide components of the thylakoid membrane system are synthesized in *Euglena* chloroplasts.

An alternative method for studying chloroplast protein synthesis makes use of the yellow mutant, y-1, of *Chlamydomonas reinhardtii.* This strain is distinguished from the wild type by an inability to synthesize chlorophyll or thylakoid membranes in the dark; after several divisions the cells become yellow. However, they retain chlorophyll precursor molecules and rapidly regenerate both the pigment and thylakoid membranes when returned to the light. Such exposure, in the presence of an inhibitor of cytoplasmic protein synthesis, e.g. cycloheximide, and labelled amino acids results in the radio-labelling of those thylakoid membrane proteins that are synthesized in the chloroplast. Polyacrylamide gel electrophoresis showed that 20 thylakoid membrane proteins were labelled.

A relatively large number of polypeptides are encoded by the chloroplast genome and synthesized on chloroplast ribosomes but identification of individual polypeptides has proved difficult. In the case of abundant chloroplast proteins, labelling experiments can be performed in isolated chloroplasts or in regreening *Chlamydomonas* y-1 and individual proteins can then be purified and tested for the presence or absence of label. This procedure has proved successful with the commonest chloroplast protein, ribulose-1,5-bisphosphate carboxylase (RuBPCase), which is a multimer of 16 subunits, eight large (molecular weight 53,000) and eight small (MW 12,000 to 15,000). Only the large subunit is synthesized in the chloroplast with the small subunits being imported from the cytoplasm.

Biochemical examination of a strain carrying a chloroplast marker can identify the translation product affected by the mutation. Antibiotic-resistant mutants of *Chlamydomonas* that lack particular ribosomal proteins have already been described and a further example is provided by mutants of *Chlamydomonas* resistant to the herbicides atrazine, bromacil and diuron. These have been shown to possess altered polypeptides associated with Photosystem II.

Identification of chloroplast translation products has been most successful with higher plants, in which some fifteen individual proteins have been assigned to chloroplast genes. These include the RuBPCase large subunit, components of the ATP synthase and coupling factor complexes, cytochrome proteins from the photosynthetic electron transport chain, ribosomal proteins and protein synthesis elongation factors.

The molecular basis of uniparental inheritance

Chlamydomonas, like all eukaryotes, possesses mitochondria as well as the chloroplast, and these mitochondria also contain an extranuclear genetic system. We must therefore consider what evidence there is that the linkage maps developed by Sager and Gillham are equivalent to the chloroplast genome and have no relevance to mitochondrial genes. Several lines of evidence suggest a chloroplast location for uniparentally inherited genes in *Chlamydomonas*, but much of this evidence is very circumstantial. Three of the more convincing arguments are (i) uniparental mutations often affect chloroplast structure and/or function (see page 44), (ii) molecular studies have shown that chloroplast DNA is physically inherited in a uniparental pattern, and (iii) a genetically-distinct set of extranuclear genes, believed to reside in the mitochondria, has been identified.

The conclusion that since uniparentally inherited mutations affect chloroplast phenotype they must reside in the chloroplast genome is rather tenuous. The location of a phenotypic effect is not always a reliable indication of gene location. However, many uniparental mutants are either impaired in photosynthesis or have altered chloroplast ribosomes, phenotypic effects that, when coupled with our knowledge of chloroplast and mitochondrial genes in other organisms, strongly suggest that these markers are located on the chloroplast genome.

Experiments performed by several different research groups have shown that chloroplast DNA is physically transmitted to meiotic products in a uniparental fashion. Their aim was to distinguish in some way between the chloroplast DNA from the mt^+ and the mt^- gametes so that the parental origin of the chloroplast DNA in the zoospores could be determined. If the zoospores could be shown to contain chloroplast DNA derived from only the mt^+ parent, this would be strong evidence for considering uniparental markers to reside on this genome. The first attempts to distinguish the parental chloroplast DNAs involved labelling of DNA with either radioactive or heavy isotopes. These experiments indicated that chloroplast DNA from the mt^- parent was degraded more rapidly than that from the mt^+ parent after gamete fusion, but that both types of parental chloroplast DNA were extensively broken down. Problems were encountered since label from degraded parental chloroplast DNA was re-utilized in later DNA synthesis, thereby complicating analysis of the results.

More convincing demonstrations of the uniparental inheritance of chloroplast DNA have been provided by two novel approaches. Firstly, the fate of chloroplast DNA from the two parents during zygote formation has been observed under the microscope using cells of *Chlamydomonas* labelled with the fluorescent dye 4′-6-diamidino-2-phenylindole (DAPI) which binds to double-stranded DNA. Rapid loss of chloroplast DNA from one of the gametes was observed, starting about 9 hours after the initial fusion event. In the second approach, crosses between two closely related species, *Chlamydomonas moewusii* and

Chlamydomonas eugametos, were studied. The chloroplast DNAs of these two species give distinct fragment patterns when digested with restriction endonucleases and so the origin of the chloroplast DNA present in the progeny can readily be determined. These crosses demonstrated that the daughter chloroplast DNA was derived exclusively from the mt^{+} parent.

Sager has proposed that this degradation of the mt^{-} chloroplast DNA is due to a restriction and modification system similar to that by which bacteria protect themselves from 'phage attack. She suggested that the chloroplast DNA in the mt^{+} gamete is extensively methylated and therefore protected from the action of the zygote-specific endonuclease which degrades the non-methylated mt^{-} chloroplast DNA. There is substantial experimental evidence for such a system and two of the nuclear genes controlling restriction and modification have been identified.

This positive evidence that uniparentally inherited markers are located on the chloroplast genome is reinforced by the presence of a separate class of extranuclear mutations that affect typical mitochondrial proteins. The important feature of these mutations is that they are not inherited in a predominantly uniparental fashion, but display a mixed pattern of inheritance. It is reasonably certain therefore, that the cytoplasmic markers discussed in the previous sections of this chapter, and placed on the linkage maps shown in Figure 19, reside on the chloroplast genome.

Molecular mapping of the chloroplast genome

Recombinant DNA technology, developed during the last decade, allows us to design and perform experiments that would previously have been impossible. This is particularly relevant to chloroplast genetics as these techniques have allowed genes to be identified on the chloroplast DNAs of organisms for which genetic analysis of extranuclear markers has proved difficult or impossible. We now know more about the location and structure of genes on the chloroplast DNA of *Euglena gracilis*, an organism with no known sexual cycle, than we do about *Chlamydomonas*, the organism chosen for the classical studies of chloroplast genetics.

Physical maps using restriction enzyme digestion have been constructed for several algal chloroplast genomes, including *Chlamydomonas* and *Euglena*, and these provide the reference points for molecular mapping of individual genes. For both these organisms the first genes to be assigned positions on the physical map were those coding for the rRNA molecules. Ribosomal RNA is abundant in the chloroplast and its preparation in bulk is relatively easy. This provides material that can be hybridized to restriction digests of chloroplast DNA to assign rRNA genes to individual fragments. These experiments revealed a fundamental difference in the organization of the chloroplast rRNA genes in *Chlamydomonas* and *Euglena*. In *Chlamydomonas reinhardtii*, as in most higher plants, these rRNA genes are arranged in two inverted repeats, each about 19 kb in length (Figure 20a). There are thus two copies of each rRNA gene, one per repeat, and each cluster has the gene order 5′-16S-7S-3S-23S-5S-3′. DNA sequencing has shown that the 23S rRNA gene contains an 870 bp intron near the 3′ end, the existence of which suggests a function for the 7S and 3S rRNA molecules. These small rRNAs have certain sequence similarities with the exon–intron boundaries within the 23S rRNA gene and may participate in the splicing of the precursor to

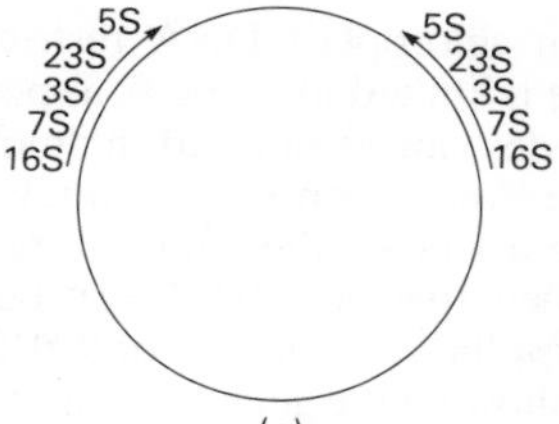

(a)

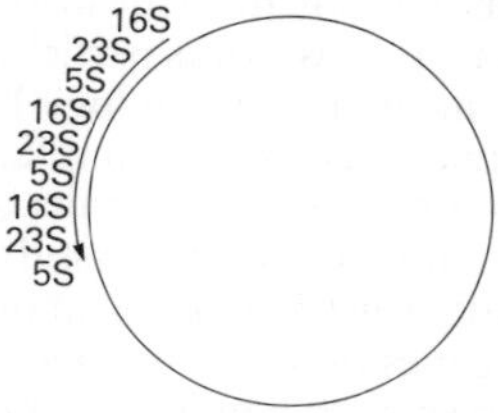

(b)

Fig. 20 Organization of algal chloroplast genes. (a) *Chlamydomonas reinhardtii*; (b) *Euglena gracilis*.

the 23S molecule in a manner analogous to the small nuclear RNA molecules (snRNAs).

The chloroplast rRNA genes of *Euglena gracilis* are arranged as three head-to-tail tandem repeats, with the gene order 5′-16S-23S-5S-16S-23S-etc.-3′ (Figure 20b). It seems that at least three copies of each rRNA gene are needed to satisfy transcriptional requirements, as a strain with only two tandem repeats has impaired chloroplast function while other varieties of *E. gracilis* with as many as five rRNA repeats are perfectly normal.

DNA–RNA hybridization has also been used to locate tRNA genes on the chloroplast genomes of *Chlamydomonas* and *Euglena*, revealing nine tRNA loci on the *E. gracilis* molecule. Four of these clusters have been sequenced (Table 10). One of the clusters, consisting of the genes for the isoleucyl and alanyl tRNAs, lies between the 16S and 23S rRNA genes, an identical arrangement to that found in higher plant chloroplast DNAs and highly reminiscent of the equivalent gene clusters in *E. coli*.

Protein coding genes have been located on chloroplast genomes by three main methods. The first involves isolation of the mRNA specific for a particular protein

Table 10 Sequence organisation of tRNA gene clusters in *Euglena gracilis*

Cluster	Gene order
I	16S rRNA–$tRNA^{ile}$–$tRNA^{ala}$–23S rRNA
II	$tRNA^{val}$–$tRNA^{asn}$–$tRNA^{arg}$–$tRNA^{leu}$
III	$tRNA^{tyr}$–$tRNA^{his}$–$tRNA^{met}$–$tRNA^{trp}$–$tRNA^{glu}$–$tRNA^{gly}$
IV	$tRNA^{glu}$–$tRNA^{ser}$–$tRNA^{fmet}$–$tRNA^{gly}$–$tRNA^{thr}$

and determining to which chloroplast DNA restriction fragment this mRNA hybridizes. This technique is limited in scope as a specific mRNA can be isolated only if the protein it encodes can be purified in amounts sufficient for antibody production. If this can be done, then the antibody can be used to precipitate polysomes that are in the act of synthesizing the relevant protein, providing a fraction from which purified specific mRNA can be obtained. In this way the *Euglena* chloroplast gene for the large subunit of RuBPCase has been mapped and subsequently sequenced, showing the gene to be split.

The second way to locate a protein coding gene is to take a DNA restriction fragment, known to contain the gene in question, from the chloroplast DNA of a related organism and to use this to locate the gene in a chloroplast DNA restriction digest from the genome of interest. This works particularly well for the chloroplast genes as there is very strong nucleotide homology between equivalent protein coding genes in different organisms. For example, there is 87% nucleotide homology between the genes for the large subunits of RuBPCase in maize and *Chlamydomonas*, more than enough for successful DNA–DNA hybridization under the experimental conditions used. This technique has already been used to locate three ATPase subunit genes on *Chlamydomonas* chloroplast DNA, using restriction fragments from the spinach genome as probes, and will undoubtedly allow more genes to be identified in the future.

The final method of gene location relies on pure chance. A DNA restriction fragment that is being sequenced because it is known to contain a particular feature, such as a tRNA cluster, will often prove to contain additional reading frames that could be protein coding genes. These reading frames can be identified by looking for homologous genes in other chloroplast genomes or, as a last resort, in *E. coli*.

The use of a combination of these techniques is permitting the gradual construction of a molecular map of the *Euglena gracilis* chloroplast genome (Figure 21). The rRNA and tRNA gene clusters, genes for the large subunit of RuBPCase, elongation factor Tu and ribosomal proteins S7 and S12 have already been located on the map and additional genes will certainly be mapped during the next few years. Similarly, five protein coding genes have now been positioned on the *Chlamydomonas* genome.

The relationship between the genetic linkage maps discussed earlier and the molecular gene maps now being produced needs to be established. One rather tedious way to do this would be to sequence the chloroplast genomes of mutant strains and try to locate the sequence anomaly between the mutant and the wild type. A better approach that has been used involves restriction digests of

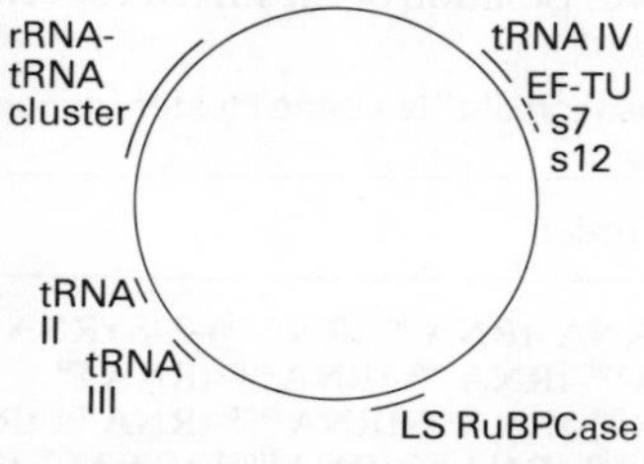

Fig. 21 Molecular map of the *Euglena gracilis* chloroplast genome.

chloroplast DNAs from *Chlamydomonas moewusii* and *Chlamydomonas eugametos*, similar to the experiment described earlier to demonstrate uniparental inheritance of chloroplast DNA in interspecific crosses. If a restriction digest is made of chloroplast DNA from a recombinant daughter cell derived from an exceptional zygote a mixture of restriction fragments will often be seen, some characteristic of one parent, some of the other. For example, in crosses between a streptomycin resistant parent and a normal strain, inheritance of a particular *Ava*-I restriction fragment was paralleled by the inheritance of the marker in recombinant cells. This suggested that the mutational event giving rise to streptomycin resistance is located within this *Ava*-I fragment. The subseqent demonstration that the DNA fragment hybridized to *Chlamydomonas* 16S rRNA indicates that the streptomycin resistance marker lies within the gene for this molecule, one of the first correlations between the genetic and molecular chloroplast maps.

Summary

Our understanding of chloroplast genetics is less advanced than that of mitochondrial genetics. The reasons for this include the greater size and complexity of the chloroplast genome, the need to distinguish chloroplast markers from mitochondrial ones and the smaller number of research workers investigating chloroplast inheritance. The principal microorganism used to study chloroplast genetics is the unicellular green alga, *Chlamydomonas reinhardtii.* Its chloroplast markers are inherited in a uniparental fashion; the alleles donated by the mt^+ parent predominate in the meiotic products following a sexual cross. The chloroplast DNA from the mt^- parent is thought to be degraded by an endonuclease synthesized in the zygote. Chloroplast DNA from the mt^+ parent is protected from degradation by being methylated. Biparental zygotes, in which the chloroplast genomes from both parents survive, arise occasionally and their frequency may be increased by artificial means. A number of methods have been devised to analyse chloroplast genetic recombination in these zygotes and in the haploid cell lines derived from their meiotic progeny. None of these methods is perfect as all involve some simplifying assumptions about the recombination process. Nevertheless, a map giving a consistent order for the chloroplast genes of *Chlamydomonas* has been constructed and recombinant DNA techniques now permit its correlation with the physical map of chloroplast DNA.

The gene expression mechanisms of chloroplasts, like those of mitochondria, show a number of prokaryotic features but, in contrast to mitochondria, the chloroplast genome employs the standard genetic code. Although large, chloroplast genomes are incapable of encoding all the proteins found in the organelle. Chloroplast DNA has been shown to encode rRNA and tRNA molecules and a number of thylakoid membrane and stromal proteins, including the large subunit of ribulosebisphosphate carboxylase, the protein synthesis elongation factor Tu, and the ribosomal proteins S7 and S12.

References

CHIBA, Y. (1951). Cytochemical studies on chloroplasts; 1. Cytologic demonstration of nucleic acids in chloroplasts. *Cytologia* 16: 259–64.

LEMIEUX, C., TURMEL, M., SELIGNY, V. L. and LEE, R. W. (1984). Chloroplast DNA recombination in interspecific hybrids of *Chlamydomonas*: linkage between a non-mendelian locus for streptomycin resistance and restriction fragments coding for 16S rRNA. *Proc. natl. Acad. Sci. U.S.A.* 81: 1164–8.

LEWIN, R. A. (ed.) (1976). The Genetics of Algae. Botanical Monographs vol. 12. Blackwell, Oxford.

SAGER, R. (1972). Cytoplasmic Genes and Organelles. Academic Press, New York.

SAGER, R. and GRABOWY, C. (1983). Differential methylation of chloroplast DNA regulates maternal inheritance in *Chlamyomonas*. *Proc. natl. Acad. Sci. U.S.A.* 80: 3025–9.

4 Miscellany

There are extrachromosomal genes or genomes which have not been assigned to a cytoplasmic organelle. These range from degenerate bacteria and viruses associated with eukaryotic cells to DNA molecules which have no known impact on the phenotype of the cell which harbours them.

The killer system in Paramecium

A number of protozoa maintain symbiotic bacteria or algae within their cytoplasm and the presence of these symbionts may alter the phenotypic properties of the host cell. Few of these associations have been examined genetically but knowledge of them is important when discussing theories of organelle evolution. The one class of associations which has been intensively studied is that between *Paramecium aurelia* and its bacterial endosymbionts.

Certain strains of *P. aurelia* have the capacity to kill other strains of the same species. In a mixed culture the killer cells release toxic particles into the medium which are ingested by the sensitive cells. The production of killer toxin is determined by the presence of distinctive particles in the cytoplasm of the killer cells. There are a number of different types of killer particles designated κ, λ, μ, γ, δ, η and σ. The various types are distinguished from one another by morphology and their effects on sensitive cells. These particles were recognized as the cytoplasmic genetic determinants of the killer phenotype but their exact relationship to the *Paramecium* cell is not clear. Biochemical studies on isolated particles have shown that they are, in fact, degenerate bacteria.

Genetics of the killer phenomenon

Most of the genetic studies have been carried out on the kappa particles and were pioneered by Sonneborn. The inheritance of killing ability due to kappa is classically cytoplasmic but the ability of the host *Paramecium* to maintain the kappa particles in its cytoplasm is determined by the possession of the dominant allele, *K*, of a nuclear gene. The effect this has on the pattern of inheritance is shown in Figure 22.

The extent of cytoplasmic exchange during conjugation in *Paramecium* depends upon the duration of mating. If the mating partners are separated soon after conjugation no cytoplasm is exchanged whereas if cell contact is extended there is cytoplasmic mixing (see Figure 3a, Chapter 1). When cytoplasmic exchange is permitted the killer character behaves like the dominant allele of a nuclear gene and all exconjugants are killers. When cytoplasmic, but not nuclear, exchange is prevented, half the exconjugants are sensitives; this shows that the killer determinant is located in the cytoplasm (Figure 22). When the exconjugants of the two different types of cross are allowed to undergo autogamy, then the first

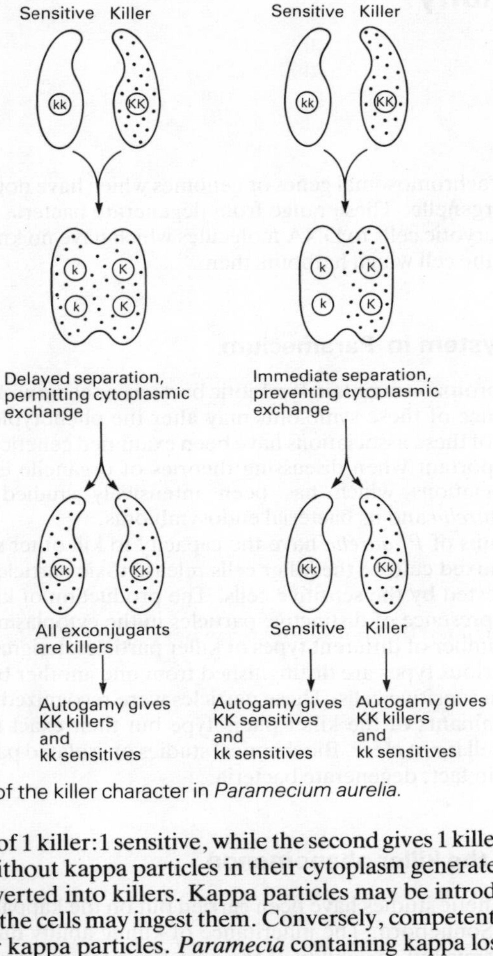

Fig. 22 Inheritance of the killer character in *Paramecium aurelia.*

cross gives a ratio of 1 killer:1 sensitive, while the second gives 1 killer:3 sensitives.

The *KK* cells without kappa particles in their cytoplasm generated in the above cross may be converted into killers. Kappa particles may be introduced by direct microinjection or the cells may ingest them. Conversely, competent killer cells can be 'cured' of their kappa particles. *Paramecia* containing kappa lose the particles when cultured in conditions favouring very high rates of cell division. It appears that the maximum rate of replication of the kappa particles is less than that of their host cell so the particles are diluted out of the cytoplasm. Kappa particles can also be cured by a number of physical and chemical agents including ultraviolet light, X-rays, high or low temperatures, the nitrogen mustards and inhibitors of protein synthesis such as streptomycin.

The results of these curing experiments forced researchers to investigate the physical nature of kappa which previously had been thought of only in rather abstract genetic terms. In particular, from X-ray curing studies, it was calculated that its X-ray sensitive volume was very large. This result prompted an intensive microscopic study of *Paramecium* cytoplasm and led to the discovery of kappa particles in killer strains.

The large size of kappa and the other killer particles and the fact that they could be cured by growth rate dilution stimulated Preer and his colleagues to suggest that they were degenerate bacteria. This has now been amply confirmed. The killer particles contain DNA, RNA, protein, carbohydrate and lipid in amounts and proportions that are typical of bacteria. The wall of the μ-particle contains diaminopimelic acid which is a uniquely bacterial feature. Killer particle rRNA has been isolated and shown to hybridize with *E. coli*, but not with *Paramecium* DNA. The DNAs of the killer particles differ from those found in both the nucleus and mitochondria of the host cells. Killer DNAs have densities in the range 1.694 to 1.708 g cm^{-3} whereas nuclear DNAs have a density range of 1.685 to 1.693 g cm^{-3}. Further, the base composition of the particles' DNA differs from both the nuclear and mitochondrial DNA of the host.

The ultrastructure of the killer particles is typically prokaryotic. They are bound by a plasma membrane and a wall and have shapes familiar to the bacteriologists—rods, cocco-bacilli and spiral rods. Particles λ, σ and δ all carry flagella. Particle α is distinguished from the others in that it inhabits the macronucleus and not the cytoplasm; it does not confer a killer phenotype on its host cell. It resembles the cellulose-degrading myxobacterium *Cytophaga*. New bacterial genera have been established for the other particles, e.g. *Caedobacter* for κ.

The knowledge that the killer particles of *Paramecium* were endosymbiotic bacteria stimulated a study of their metabolism and attempts were made to culture them outside the host cell. Isolated particles are able to respire glucose and contain the enzymes of both the tricarboxylic acid cycle and the pentose phosphate shunt. Kappa and μ have a cytochrome spectrum which differs from that of the host.

It has been claimed that both λ and μ can be cultured outside the host cell. They divide very slowly, once every 24 hours, and can achieve a maximum population density of only 200,000 particles per ml. The λ and μ particles cultured in this way are said to kill sensitive cells. This last result has led people to question these data. Mu particles usually only produce their killing effect when a μ-killer cell is mated with a sensitive cell. When μ particles are isolated from *Paramecium* they do not kill sensitive cells.

The endosymbionts of *Paramecium* do not all have killing ability. The α particles of the macronucleus are not killers and kappa particles can apparently mutate to a non-killer form known as π. These π particles differ from wild-type kappa as they are unable to form R-bodies. These refractile structures, consisting of a ribbon of protein rolled up on itself, are found in the cytoplasm of between 1 and 40% of kappa particles. The higher proportion is seen when the host cells are dividing slowly. Kappa particles which contain R-bodies are highly refractile when examined by phase contrast microscopy and are known as 'brights'. The killing activity of kappa particles is confined to these brights. However, it is only the non-brights which can reproduce and which are infective. It is probable that the R-body itself is the killer toxin.

The R-bodies have virus-like particles associated with them. These are hexagonal structures, between 15 and 120 nm in diameter, and some of them have tails. These virus-like particles or phages have been isolated. They contain both protein and DNA, the latter having a buoyant density which is distinct from that of the kappa particle itself. The amount of DNA in the virus-like particle is similar to that found in a T-even phage. These phages are rare or absent in non-bright kappa particles and it has been suggested that they exist there in a prophage state and that they may be induced to replicate themselves and elaborate the R-body. It is certainly true that the production of R-bodies may be induced by the u.v. irradiation of kappa particles and u.v. light is known to induce prophages. It is not unusual for bacterial toxins to be specified by phages; the diphtheria and botulism toxins are examples.

Some strains of *Paramecium aurelia* have the ability to kill other, sensitive strains. This trait was found to be inherited cytoplasmically but to be dependent on a number of maintenance genes in the nucleus. *Paramecia* can be cured of the killer character by treatment with a variety of physical and chemical agents or by culture at very high growth rates. The killer character correlates with the presence, in the cytoplasm, of large particles which look like bacteria. Chemical and biochemical studies have demonstrated that the killer particles are indeed degenerate bacteria. The ability of these bacteria to bestow killing activity on their host is dependent on the production of a protein toxin which appears to be specified by a bacteriophage found within the killer particles. These relationships are summarized in Figure 23.

The killer phenomenon and virus-like particles in fungi

Fungi which have the ability to kill other members of the same species are not uncommon. The phenomenon is associated with the presence of defective viruses. The genomes of these virus-like particles (VLPs) consist of one or more molecules of double-stranded RNA. The genetic and molecular basis of killer ability in fungi has been studied most thoroughly in *Saccharomyces cerevisiae* and in the basidiomycete, *Ustilago maydis*, the organism which causes corn smut.

Killer yeast Many laboratory strains of *S. cerevisiae* have the ability to kill other strains of the same species, known as sensitives. Killing takes place at low pH, 4.2 to 4.8, and is due to the disruption of the membrane of the sensitive cell by a protein toxin of molecular weight 11,000 which is released from the killer cells. The toxin binds tightly to the wall of the sensitive cell and causes the leakage of

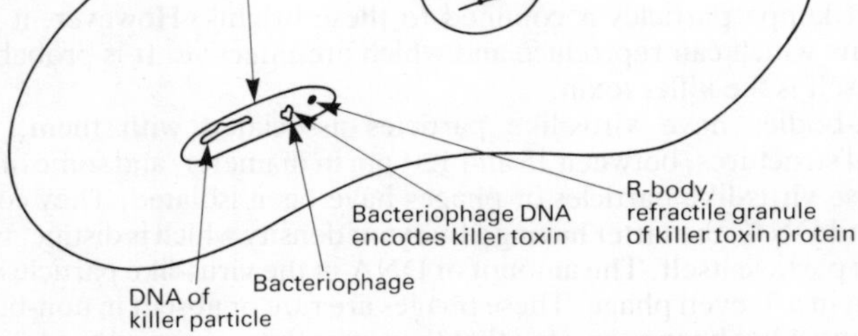

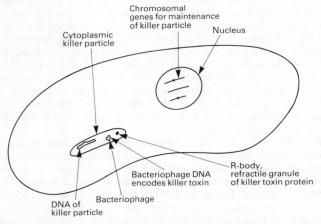

Fig. 23 Relationships between nuclear and cytoplasmic genomes in killer *Paramecium*.

intracellular K^+ and ATP through the membrane. The sensitive cell dies since it no longer has an energy supply which can sustain macronuclear synthesis and other metabolic activities. Sensitive strains are comparatively rare in laboratory collections of *S. cerevisiae* but most commercial strains are sensitives. There have been a few reports of fermentations in breweries being spoilt due to infection with killer yeast and one saké yeast strain has been deliberately converted into a killer to prevent such occurrences.

The killer phenotype in yeast shows a non-Mendelian 4:0 segregation pattern in meiotic tetrads so its genetic determinant must be extrachromosomal. The killer genome can be shown to be cytoplasmically located using the heterokaryon test. This test is made, in yeast, by a so-called cytoduction experiment in which cytoplasmic material is exchanged between two mating cells without their nuclei fusing. Nuclear fusion is prevented by having one of the pair carry a mutation in the *kar* or karyogamy gene.

The killer phenomenon was first discovered in Bevan's laboratory and it was his group which demonstrated the association of killing with double-stranded RNA (dsRNA). Killer cells contain two types of dsRNA molecule, P1 or L dsRNA which has a molecular weight of 3.4×10^6 and P2 or M dsRNA, molecular weight $1.6–2.0 \times 10^6$. Figure 24 shows an electron microscope photograph of linear L dsRNA with a circular phage DNA as a size standard. The L and M molecules do not have any sequences in common. The L molecule appears to be present in almost all laboratory strains of *S. cerevisiae* and many of these also contain M and are therefore killers. Double-stranded RNA is rare in commercial brewing strains.

Strains which contain no dsRNA at all or which have only the L and not the M molecule are sensitives. Other strains occur, called neutrals, which are immune to the killer toxin but cannot produce that toxin themselves. Neutral strains appear to have either point or deletion mutations in M dsRNA. This leads to the conclusion that the M dsRNA is the genome which specifies both toxin production and immunity. The relationship between the different phenotypes and the dsRNA content of the cells is summarized in Table 11.

These dsRNA molecules are associated with VLPs within the yeast cell. The particles are polygonal and have a diameter of 40 nm (see Figure 25). The L dsRNA is the genome which specifies these VLPs; L dsRNA has been translated *in vitro* to produce the major capsid protein of the VLPs which has a molecular weight of about 75,000. Both L and M dsRNAs may be found encapsulated in the VLPs and the particles contain RNA polymerases which are thought to be responsible for the replication and expression of the dsRNA genomes.

Table 11 The relationship between dsRNA content and killer phenotype in yeast

Cell type	Genotype	dsRNAs present	Phenotype
Killer	[KIL-k]	L + M	Kills Resistant to killing
Sensitive	[KIL-o]	None or only L	Does not kill Killed by killers
Neutral	[KIL-n]	L + mutant M	Does not kill Resistant to killing

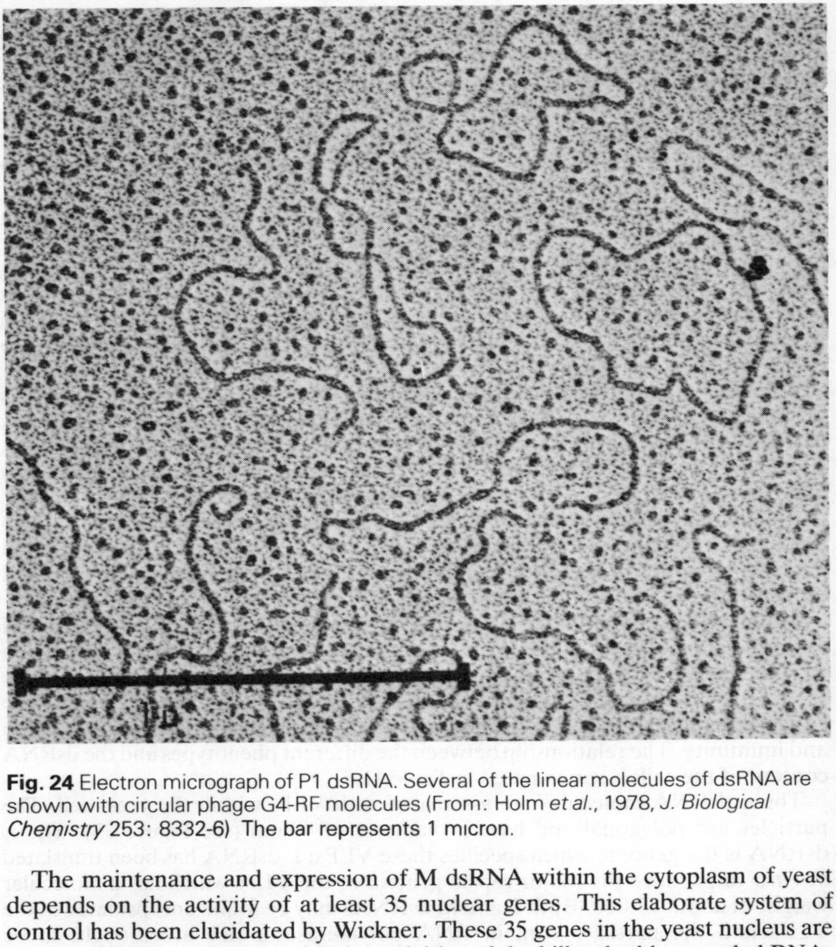

Fig. 24 Electron micrograph of P1 dsRNA. Several of the linear molecules of dsRNA are shown with circular phage G4-RF molecules (From: Holm *et al.*, 1978, *J. Biological Chemistry* 253: 8332-6) The bar represents 1 micron.

The maintenance and expression of M dsRNA within the cytoplasm of yeast depends on the activity of at least 35 nuclear genes. This elaborate system of control has been elucidated by Wickner. These 35 genes in the yeast nucleus are not specifically concerned with the activities of the killer double-stranded RNA; many of them encode normal cell functions for which the killer dsRNA has a particularly stringent requirement. For instance, mutants of the maintenance genes *pet*-18, *mak*-1 and *mak*-16 are temperature sensitive for cell growth and division. Mutants of *pet*-18 are unable to maintain either M dsRNA or mitDNA, although they have no effect on other yeast extrachromosomal elements. The *kex*-2 gene, required for the expression of the killer phenotype, is also involved in mating and sporulation. Many of the other genes involved in killer maintenance and expression have less well-defined effects on cell growth, producing small colonies on agar plates. The study of these genes is a valuable method of investigating the requirements for dsRNA replication and function and presents a model system for the study of virus-host interactions.

Killer in *Ustilago maydis* The killer system in *Ustilago maydis* is much more

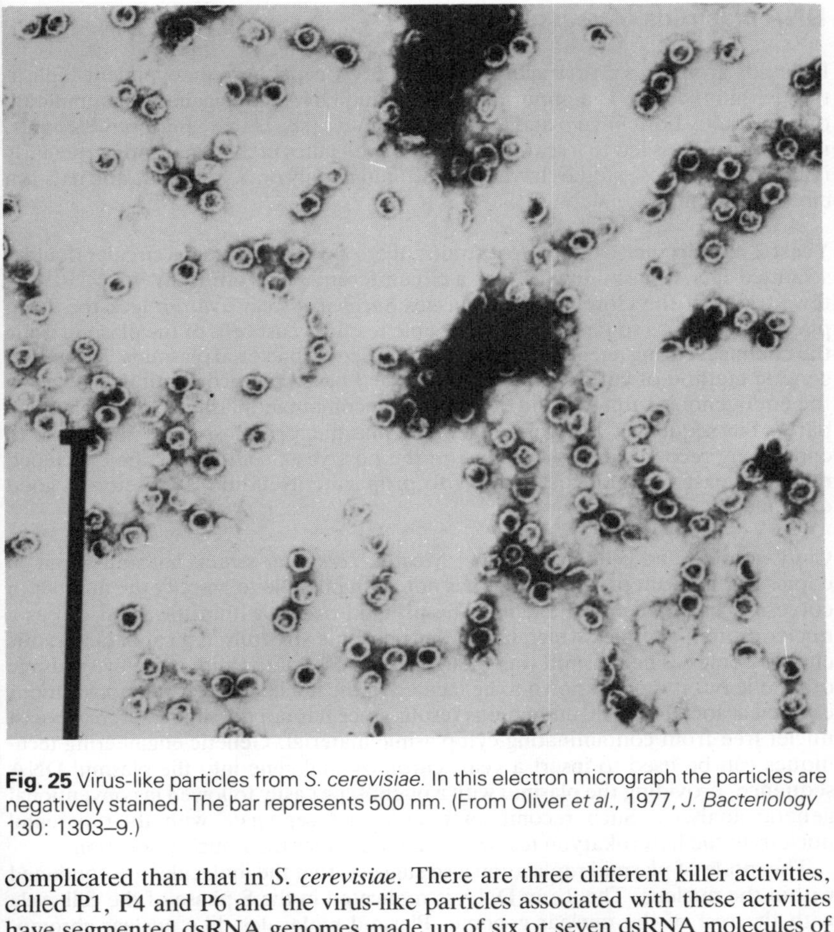

Fig. 25 Virus-like particles from *S. cerevisiae*. In this electron micrograph the particles are negatively stained. The bar represents 500 nm. (From Oliver *et al.*, 1977, *J. Bacteriology* 130: 1303–9.)

complicated than that in *S. cerevisiae*. There are three different killer activities, called P1, P4 and P6 and the virus-like particles associated with these activities have segmented dsRNA genomes made up of six or seven dsRNA molecules of different sizes. Segmented genomes are common among the dsRNA viruses of the filamentous fungi.

As with *Saccharomyces* and *Paramecium* a given killer strain is immune to its own toxin but is sensitive to the two other types of toxin. Thus, a P1 killer is resistant to the P1 toxin but may be killed by the toxin of P4 and P6 strains. *Ustilago* is remarkable in that dsRNA molecules may be maintained in the cytoplasm without being encapsulated in a VLP. This situation is found in some suppressive sensitive strains which prevent the replication of dsRNA associated with P4 and P6 killing activity. These cells contain two or six different molecules of dsRNA but do not appear to contain VLPs.

DNA plasmids of yeasts

Plasmids are pieces of extrachromosomal DNA capable of independent replication within the cell. The importance of plasmid DNA in the genetic manipulation of bacteria by both *in vivo* and *in vitro* techniques (see Hardy: *Bacterial Plasmids*, in this series) has led to a search for similar molecules in the fungi and particularly in the yeasts. These studies have concentrated mainly on *S. cerevisiae* and its 2 μm circular plasmid.

Yeast 2 μm **circular DNA** Most strains of *S. cerevisiae* contain a circular double-stranded DNA plasmid which has a circumference of 2 μm (MW 4.1×10^6). A few strains of the closely related species *Saccharomyces uvarum* lack the 2 μm plasmid but it has so far proved impossible to cure yeast cells of the plasmid using the chemical curing agents which have been used on bacterial plasmids. Recently a genetic method of curing has been developed based on incompatibility between the endogenous 2 μm plasmid of yeast and recombinant plasmids carying all, or a part, of its sequence. The possesion of the plasmid, or its loss, does not appear to confer any recognizable phenotype on the host yeast cell. It must be concluded that the yeast plasmid exists merely to propagate itself and is therefore a good example of a selfish gene.

Copy number and cellular location Most *S. cerevisiae* strains harbour about 50 copies of the 2 μm plasmid and it has not been possible to amplify the number of copies by using chemical agents. The plasmid occurs within the yeast cell as a nucleoprotein complex which has the nucleosome structure typical of eukaryotic chromosomes. The plasmid was thought to be associated with some cytoplasmic organelle but it is now known to be located in the nucleus. Direct demonstrations of nuclear location yield ambiguous results since it is not possible to prepare yeast nuclei free from contaminating cytoplasmic material. Genetic engineering techniques can be used to insert a yeast chromosomal gene into the plasmid DNA sequence, providing the plasmid with a phenotype easily followed in conventional genetic analyses. Such recombinant molecules segregate with their parental nucleus in the heterokaryon test which demonstrates their nuclear location.

The mode of plasmid replication also suggests that the molecules are contained within the nucleus. The 2 μm DNA is replicated in the S-phase of the cell cycle with the rest of the nuclear genome. Plasmid molecules, like nuclear chromosomes, are replicated only once per S-phase.

Molecular structure The structure of the yeast 2 μm plasmid has been examined in detail using restriction endonuclease analysis and its complete nucleotide sequence has been determined by Hartley and Donelson (1980). A map of the plasmid is shown in Figure 26. Two regions of the plasmid contain exactly the same nucleotide sequence which are inverted with respect to one another. These are separated by two unique sequences of unequal length. The homology of the two inverted repeats permits intramolecular recombination to take place between them and generates two forms of the 2 μm molecule called A and B (Figure 26). Plasmid DNA purified from a given strain of yeast is a mixture of these A and B forms.

Plasmid gene expression Gene expression by the 2 μm plasmid has been studied

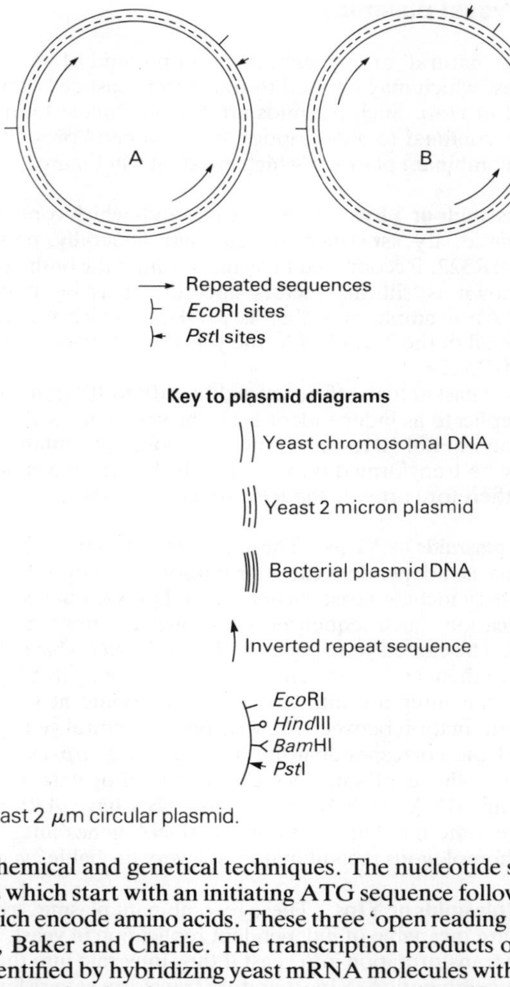

Fig. 26 The yeast 2 μm circular plasmid.

by both biochemical and genetical techniques. The nucleotide sequence contains three regions which start with an initiating ATG sequence followed by a sequence of triplets which encode amino acids. These three 'open reading frames' have been named Able, Baker and Charlie. The transcription products of these sequences have been identified by hybridizing yeast mRNA molecules with 2 μm DNA. The polypeptides encoded by these messages have been identified using an *in vitro* translation system. It appears that both strands of the entire circular DNA molecule are transcribed and that different transcripts are produced from the A and B forms.

A functional analysis of the different regions of the yeast plasmid has been carried out by making mutants *in vitro* using recombinant DNA techniques. Three functional genes have been specified: *FLP* which is contained within the Able region and is required for the 'flip' function of intramolecular recombination and *REP*-1 and *REP*-2, which correspond to the Baker and Charlie regions. They are required for the replication of the plasmid. The replication origin itself has been mapped to a region spanning a portion of one inverted repeat and part of the adjacent unique sequence.

Recombinant yeast plasmids

In addition to the 'natural' or endogenous 2 μm plasmid of yeast, a number of plasmid molecules, which may be used to transform yeast cells genetically, have been constructed *in vitro*. Such plasmids are literally 'made to measure' so the discussion will be confined to a description of the general properties of the five classes of such recombinant plasmids which are shown in Figure 27 a to e.

Yeast episomal plasmids or YEps These are plasmids which contain all or part of the yeast 2 μm circle, a yeast structural gene and, generally, part of an *E. coli* plasmid such as pBR322. Recombinant plasmids containing both yeast and *E. coli* sequences are known as 'shuttle' vectors since they may be used to transform either organism. An example of a YEp is pJDB219 which was constructed by Beggs. It contains all of the 2 μm DNA, the yeast structural gene *LEU*-2 and the *E. coli* plasmid pBR322.

YEps transform yeast at high efficiency yielding 10^3 to 10^5 transformants per μg of DNA. They replicate as independent extrachromosomes within the yeast cell and have a copy number of 50 to 100. If the YEp does not contain all of the 2 μm DNA it may only be transformed into strains which carry the endogenous 2 μm plasmid and can therefore provide the *REP* functions in *trans*.

Yeast replicative plasmids or YRps These plasmids also transform yeast at high efficiency and exist as independent extrachromosomes. They are able to replicate in yeast because they include yeast chromosomal DNA sequences which contain an origin of replication. Such sequences are known as autonomously replicating segments or *ARS*. The cloned yeast genes *TRP*-1 and *ARG*-4 have *ARS* sequences located adjacent to them and these genes are often included in YRp plasmids.

YRp plasmids can integrate into a yeast chromosome at low frequency by homologous recombination between the wild type structural gene on the plasmid (e.g. *TRP*-1) and the corresponding mutant gene (e.g. *trp*-1) in the host. An extreme example of this is plasmids which contain all or part of the yeast gene encoding ribosomal RNA (rDNA). Such sequences have *ARS* properties but since the yeast genome has 140 copies of the rDNA genes integration into the chromosome by homologous recombination is a very probable event.

Yeast integrative plasmids or YIps These recombinant plasmids contain no *ARS* sequence and so are incapable of independent replication in yeast. They can only survive following transformation into yeast if they integrate into the chromosome by homologous recombination. YIps therefore transform at very low frequency (1 to 10 transformants per μg of DNA) but once integrated they are extremely stable.

Yeast centromere plasmids or YCps YCps are recombinant plasmids with a cloned fragment of yeast DNA which contains the DNA sequence of the centromere of a chromosome. The centromeres of chromosome III (*CEN*-3) and chromosome XI (*CEN*-11) have been cloned by Carbon's group. Most YCp plasmids also contain an *ARS* sequence and are therefore able to replicate independently in yeast. They behave as mini-chromosomes segregating 2:2 at meiosis whereas other types of recombinant plasmids and the endogenous 2 μm circles give the 4:0 ratio characteristic of non-Mendelian inheritance.

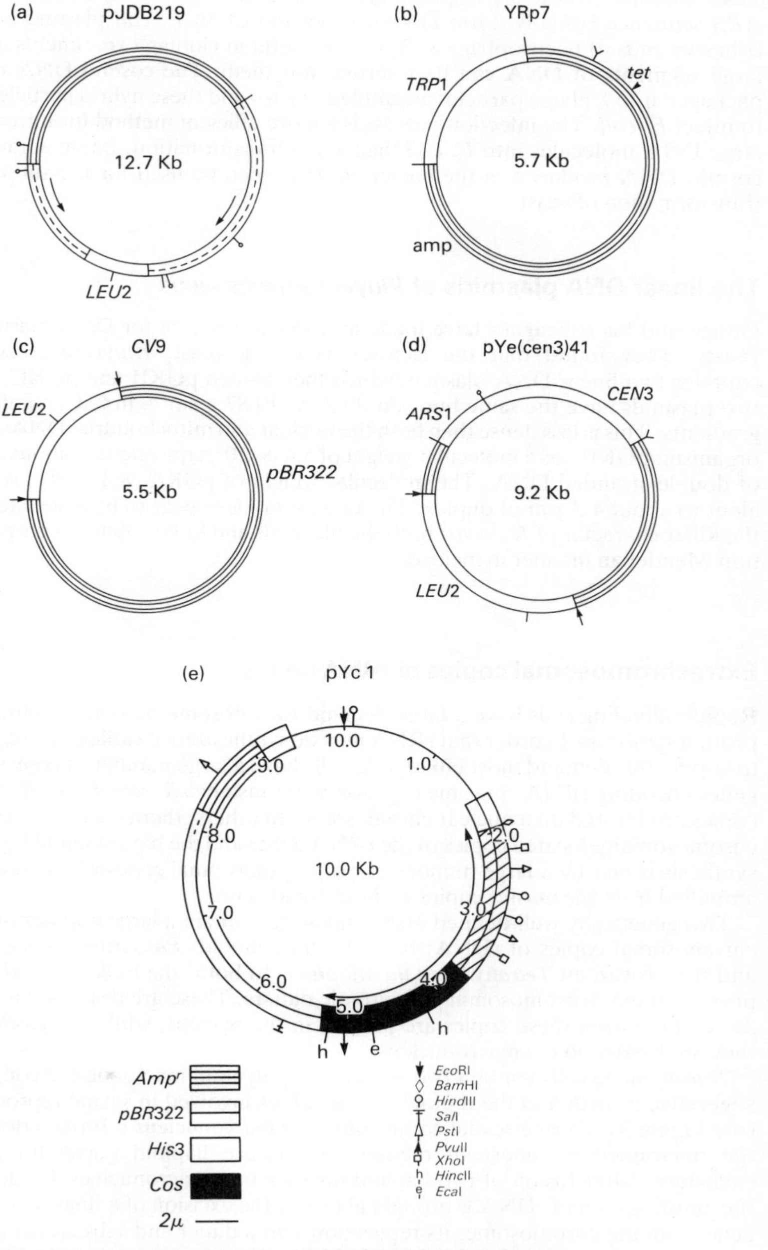

Fig. 27 (a) *A YEp*; (b) *A YRp*; (c) *A YIp*; (d) *A YCp*; (e) A Yeast Cosmid.

Yeast cosmids These are a special class of shuttle vector which contain a yeast *ARS* sequence (usually 2 μm DNA), a portion of an *E. coli* plasmid and the cohesive ends of bactriophage λ. They are useful in cloning experiments as very large segments of DNA can be inserted into them. The cosmid DNA can be packaged into λ phage particles assembled *in vitro* and these hybrid particles used to infect *E. coli*. The infection process is a more efficient method for introducing large DNA molecules into *E. coli* than using transformation. Large amounts of cosmid DNA produced in the bacterium may then be used for the subsequent transformation of yeast.

The linear DNA plasmids of *Kluyveromyces lactis*

Gunge and his colleagues have made an extensive search for DNA plasmids in yeasts. They found that the lactose-fermenting yeast, *Kluyveromyces lactis* contains two linear DNA plasmics which they named pGKl1 and pGKl2. These two plasmids have the same buoyant density, 1.687 g cm^{-3}, in CsCl equilibrium gradients. This is less dense than both the nuclear and mitochondrial DNAs of this organism. pGKl1 has a molecular weight of 5.4×10^6, equivalent to about 2.7 μm of double-stranded DNA. The molecular weight of pGKl2, 8.4×10^6, is equivalent to about 4.2 μm of duplex. The two plasmids appear to be associated with the killer character of *K. lactis*: both the plasmids and killing ability segregate in a non-Mendelian manner in meiosis.

Extrachromosomal copies of rRNA genes

Rapidly-dividing cells have a large demand for ribosome biosynthesis to sustain protein synthesis. In order that rRNA may be synthesized at sufficiently high rates to supply this demand most eukaryotic cells have a large number of copies of the genes encoding rRNA. In some organisms, for instance *S. cerevisiae*, all of these copies are located on a nuclear chromosome. In others, there is only one or a few chromosomally-located copies of the rRNA genes and the high demand for rRNA synthesis is met by a large number of extrachromosomal genes which have been amplified from the master copies in the chromosome.

Two genetically well-defined microorganisms contain a large number of extrachromosomal copies of rDNA: the true slime mould, *Physarum polycephalum* and the protozoan *Tetrahymena thermophila*. In both, the bulk of the rDNA is present as extrachromosomal palindromic dimers. These are described in Figure 28. In *Physarum* these copies are present in the nucleus, while in *Tetrahymena* they are located in the macronucleus.

Tetrahymena cells contain two nuclei, the polyploid macronucleus controlling vegetative growth and the diploid micronucleus involved in sexual reproduction (see Figure 3). During sexual conjugation the macronucleus is broken down and the micronucleus undergoes meiosis to produce haploid nuclei for genetic exchange. After fusion of two haploid nuclei a new macronucleus develops and the amplification of rDNA is brought about by the excision of a single copy of the gene from the chromosome, its replication into a dimer and subsequent production of numerous copies of this amplified form.

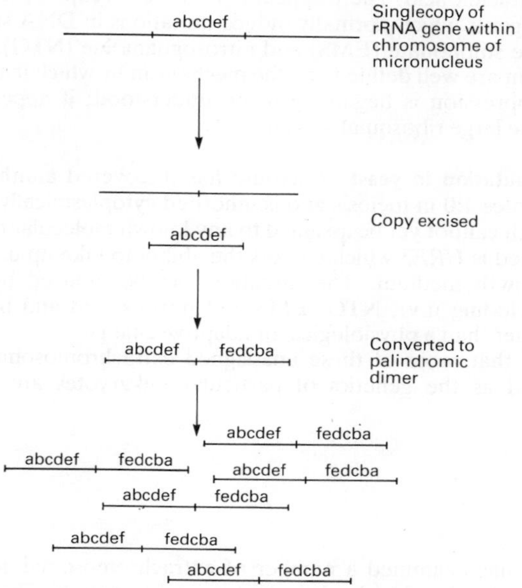

Fig. 28 Amplification of ribosomal RNA genes in *Tetrahymena thermophila.*

Genes in search of a molecule

The *psi* factor in yeast The genetic code is punctuated by so-called 'stop' codons which do not encode an amino acid and cause the ribosome to terminate the synthesis of a polypeptide chain when they encounter such a codon in an mRNA molecule. Mutations, known as 'nonsense' mutations occur which introduce a stop codon into a gene in a place where it would not naturally be found and thereby cause the premature termination of the polypeptide encoded by that gene. These nonsense mutations can be reversed or suppressed by secondary mutations in genes encoding tRNA molecules. These suppressor mutations alter the anticodon carried by the tRNA such that it now decodes the nonsense stop codon and inserts an amino acid into the growing polypeptide chain, thus preventing premature termination and restoring at least partial wild-type function to the completed protein.

The efficiency of suppression of nonsense mutations is often very low. Cox has isolated a mutant of *S. cerevisiae* which increases the efficiency of suppression. The effect is specific to mutants in which the nonsense codon is UAA ('ochre mutants'). Strains with such increased suppression efficiency are known as psi^+ and this phenotype is inherited extrachromosomally. It segregates 4:0 in meiotic tetrads and was shown to be cytoplasmically located by a heterokaryon test.

It has not been possible to assign the psi^+ phenotype to any of the known extrachromosomes of yeast: mitochondrial DNA, the 2 μm plasmid or the dsRNAs. However, indirect evidence suggests that the *psi* element is an independent nucleic acid. There appear to be about 60 copies per cell and it can be mutated by agents which normally induce mutations in DNA such as u.v. light, ethyl methane sulphonate (EMS) and nitrosoguanidine (NTG). The genetics of the *psi* element are well defined and the mechanism by which it mediates its effect on ochre suppression is beginning to be understood; it appears to affect the function of the large ribosomal subunit.

The URE3 mutation in yeast Lacroute has discovered another gene in yeast which segregates 4:0 in meiosis and is inherited cytoplasmically in heterokaryon tests but which cannot yet be assigned to any known molecular determinant. The gene concerned is *URE*3 which confers the ability to take up ureidosuccinic acid from the growth medium. The mutation can be induced by classical DNA mutagens including u.v., NTG, EMS and nitrous acid and behaves as a true mutation rather than a physiological or adaptive change.

It is likely that more of these unassigned extrachromosomal mutations will be discovered as the genetics of particular eukaryotes are studied in more detail.

Summary

This chapter has examined a number of extrachromosomal genes or elements which are not associated with mitochondria or chloroplasts. The killer particles of *Paramecium* appear to be degenerate bacteria which exist as symbionts within the cytoplasm of the protozoan. These particles confer on their host the ability to kill other *Paramecia*. The toxin involved is itself encoded by a bacteriophage contained within the killer particle. Killer fungi are also known but here the protein toxin is encoded by double-stranded RNA molecules associated with cytoplasmic virus-like particles (VLPs). No infective cycle has been demonstrated for these VLPs but their present symbiotic lifestyle may have evolved from more virulent, or parasitic, antecedents. The most perfect symbiosis is exhibited by the 2 μm nuclear plasmid DNA of yeast. This molecule contains three genes which determine its own replication and recombination. The plasmid seems to have no effect on the phenotype of its host. There are some genes, such as *psi* and *URE*-3 in yeast, which have been shown to be cytoplasmically located by genetic tests but for which no molecular determinant is known. It remains to be seen whether these are associated with some new endosymbiont or whether they represent some quirk in the genetic organization of a eukaryotic cell.

References

AIGLE, M. and LACROUTE, F. (1975). Genetic aspects of ure3, a non-mitochondrial, cytoplasmically inherited mutation in yeast. *Molec. Gen. Genet.* 136: 327–45.

BEGGS, J. D. (1981). Gene cloning in yeast. In; *Genetic Engineering* Vol. 2 (Edited by R. Williamson) pp. 175–203. Academic Press, London.

BOTSTEIN, D. and DAVIS, R. W. (1982). Principles and practise of recombinant DNA research in yeast. In: *The Molecular Biology of the Yeast* Saccharomyces*: Metabolism and Gene Expression* (Edited by J. N. Strathern, E. W. Jones and J. R. Broach) pp. 607–36. Cold Spring Harbor Laboratory, New York.

BROACH, J. R. (1981). The yeast 2 μ circle. In: *The Molecular Biology of the Yeast* Saccharomyces*: Life Cycle and Inheritance. Ibid* pp. 445–70.

BUSSEY, H. (1981). Physiology of killer factor in yeast. *Advances in Microbiological Physiology* 22: 93–122.

CLARKE, L. and CARBON, J. (1980). Isolation of a yeast centromere and construction of functional small circular chromosomes. *Nature* 287: 504–9.

COX, B. S. (1965). ψ, a cytoplasmic super-suppressor in yeast. *Heredity* 20: 505–21.

DIN, N. and ENGBERG, J. (1979). Extrachromosomal ribosomal RNA genes in *Tetrahymena.* Structure and evolution. *Journal Molecular Biology* 34: 555–74.

GUNGE, N. (1983). Yeast DNA plasmids. *Ann. Rev. Microbiol* 37: 253–276.

HARTLEY, J. L. and DONELSON, J. E. (1980). Nucleotide sequence of the yeast plasmid. *Nature* 286: 860–4.

PREER, J. R., PREER, L. B. and JARARD, A. (1974). Kappa and other endosymbionts in *Paramecium aurelia. Bacteriol. Revs* 38: 113–63.

WICKNER, R. B. (1979). The killer double stranded RNA plasmids of yeast. *Plasmid* 2: 303–22.

WICKNER, R. B. (1981). Killer systems in *Saccharomyces cerevisiae.* In; *The Molecular Biology of the Yeast* Saccharomyces*: Life Cycle and Inheritance* (Edited by J. N. Strathern, E. W. Jones and J. R. Broach) pp. 415–44. Cold Spring Harbor Laboratory, New York.

5 The evolution of eukaryotic extrachromosomes

The investigation of and speculation about the evolution of eukaryotic extrachromosomes is of great importance because such elements may be involved in the exchange and rearrangement of genes in eukaryotic organisms and, more importantly, the evolution of cytoplasmic organelles is central to theories on the evolution of the eukaryotic cell itself. Two principal hypotheses have been advanced to explain the origin of eukaryotic cells: in the *Xenogenous* or *Endosymbiont hypothesis* (Figure 29) it is envisaged that the nuclear and organellar genomes initially inhabited different kinds of cell and that prokaryotic cells were recruited into a primitive eukaryote as endosymbionts. The precursor to the chloroplast was probably a blue-green alga, while the progenitor of the mitochondrion may have been an aerobic bacterium similar to *Paracoccus*. The *Autogenous* or *'Direct filiation' hypothesis* proposes that the nuclear and organellar genomes became physically compartmentalized and functionally specialized within a single kind of cell. Two possible routes are suggested for such a process: either, at an early stage of evolution a protoeukaryote emerged which subsequently organized its genome in a number of separate cellular compartments

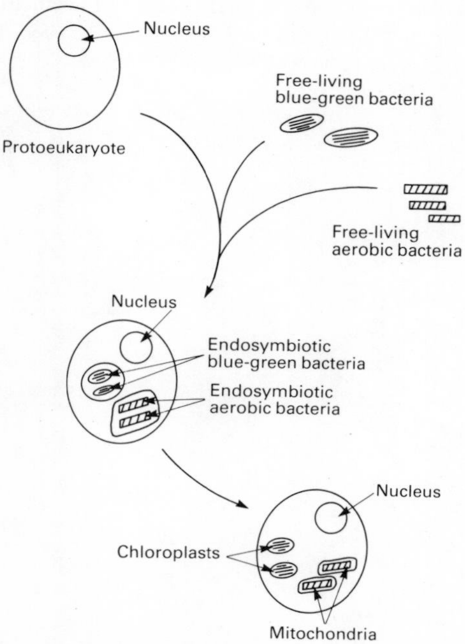

Fig. 29 Xenogenous or endosymbiont theory of organelle evolution.

(Figure 30a), or the first eukaryotes arose from some loose association or consortium of different kinds of prokaryotic cells (Figure 30b). The members of the consortium subsequently specialized to form nucleus, mitochondria and chloroplasts.

Evaluation of these competing hypotheses requires consideration of a number of points, the most important being that the prokaryotic nature of organelles is compatible with either theory even though it is usually claimed to substantiate the endosymbiotic route. Proof of either hypothesis requires the demonstration of either the separate (xenogenous) or the same (autogenous) cellular origin of the nuclear and organellar genomes. It should be emphasized that the evolution of eukaryotic cells was probably complex. For example, there is considerable evidence for the exchange of genetic information between organelles and the nucleus and between chloroplasts and mitochondria. These kinds of genetic exchange make it extremely difficult to deduce the true lineage of a particular genome.

The evolution of plasmids and virus-like particles

Plasmids and viruses are known to have contributed to bacterial evolution by promoting the exchange of genes between both closely and distantly related organisms. It is conceivable, but poorly substantiated, that the same is true of eukaryotes; while it is known that tumour viruses can be integrated into the nuclear genome of higher organisms there does not seem to be a comparable situation among the microbial eukaryotes. The only eukaryotic extrachromosomes which are known to have a duplicate of their sequences within the nuclear

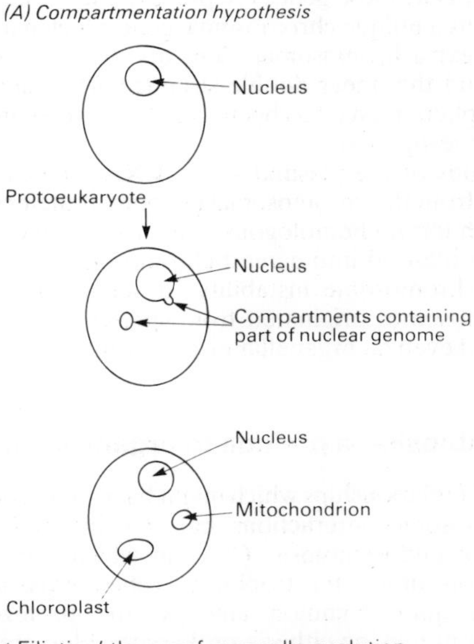

Fig. 30a Autogenous or 'Direct Filiation' theory of organelle evolution.

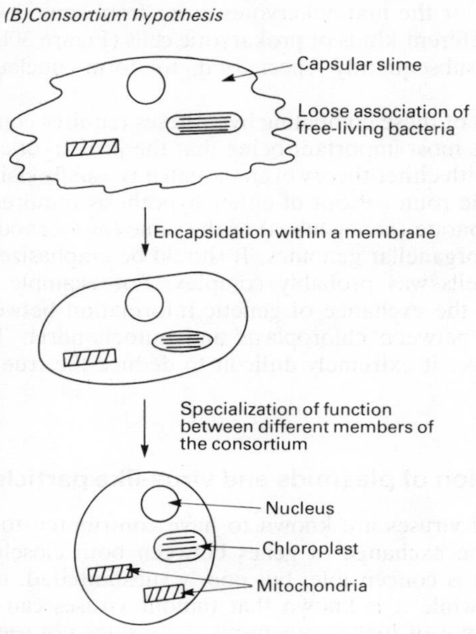

Fig. 30b Consortium hypothesis.

chromosomes are the ribosomal RNA genes. In some eukaryotic microbes, such as *Tetrahymena*, these genes are present both as a multicopy set of extrachromosomes and as a unique chromosomal gene which maintains the germ line. None of the other extrachromosomal elements discussed in Chapter 4, circular DNA plasmids and the linear double-stranded RNA and DNA molecules associated with killer phenomena, has been found to have a chromosomal gene copy, despite intensive investigation.

The origins of the plasmid and dsRNA molecules remain a mystery. If they originated from the chromosomal genome of their current hosts, they did so in an event which left no homologous sequences behind. The artificial insertion of the yeast 2 μm plasmid into a yeast chromosome by recombinant DNA techniques results in chromosome instability. It seems likely then that these eukaryotic extrachromosomes originated from species other than the ones they currently inhabit, and even an organellar origin for them cannot be excluded.

Endosymbiosis—a prelude to organelle status

The types of relationships which might have preceeded present-day chromosome–extrachromosome interactions can be inferred by studying contemporary examples of endosymbiosis. Cases in which a photosynthetic organism is the endosymbiont of a heterotrophic microbe are particularly relevant. Such a relationship is quite common amongst the protozoa and the photosynthetic endosymbiont may be either a prokaryotic blue-green bacterium or a eukaryotic alga. The cell division cycles of host and endosymbiont must be coordinated so that each daughter cell inherits approximately the same number of algae. Some protozoa contain several hundred algae per cell so this coordination need not be a very close one, but when there are very few endosymbionts it must be exact. The protozoan *Cyanophora* contains two blue-green bacterial cells in each of its cells. When the host divides each daughter cell receives one blue-green bacterium which immediately divides to restore the number of endosymbionts per cell to two. When this degree of coordination is involved the endosymbiont approaches the status of an organelle.

The principal distinction between an endosymbiont and an organelle is that the former is capable of growth and division outside the host while the latter is not. Nowhere is this distinction more blurred than in the killer particles of *Paramecium* (see Chapter 4). It is doubtful whether these particles can be cultured independently although claims for this have been made. The retention of a cell wall by the killer particles and the ability to uncouple their rate of division from that of the host indicates that these particles represent a transitional form between endosymbiont and organelle.

The existence of contemporary endosymbionts encourages the belief that organelles had an endosymbiotic origin. However, present-day eukaryotes, which harbour endosymbionts, already contain organelles and the possibility of successive symbioses further confuses understanding of their lineage. It is a distinct possibility, for example, that certain chloroplasts evolved from *eukaryotic* algal endosymbionts.

Organelle evolution

A discussion of the evolution of mitochondria and chloroplasts inevitably centres on the evidence for and against the xenogenous (endosymbiont) hypothesis. The endosymbiotic origin of chloroplasts is widely accepted, whereas that of mitochondria is more doubtful because of their eukaryotic as well as many unique characteristics. Our knowledge of the molecular biology of chloroplasts is less advanced than that of mitochondria; when we know more about them we may be less confident of their lineage.

Chloroplast evolution Organization of the chloroplast genome is in many respects similar to that of bacteria: all chloroplast DNA molecules are covalently closed circles and are free of histones. The majority have a molecular size of 40 μm to 50 μm, about the same size as the genome of a large bacteriophage.

This means that many chloroplast proteins must be encoded by the nucleus. The marine alga *Acetabularia* is an apparent exception. Renaturation kinetics have demonstrated that its chloroplast has a genome size of 750 μm which would be sufficient to encode its entire complement of proteins. This demonstrates that chloroplasts may have evolved from organisms with a complete genome capable of supporting an autonomous existence.

The organization of genes within chloroplast DNA follows the prokaryotic pattern, although certain protein-coding genes and the 23S rRNA gene of *Chlamydomonas* has been found to contain introns. Most chloroplast genomes have, like bacteria, two to three copies of the rRNA genes. There is no equivalent of the eukaryotic 5.8S rRNA and the spacer between the 16S and 23S genes has been shown to contain sequences encoding tRNA molecules; another prokaryotic feature. Studies of *Euglena* have demonstrated that the spacer contains the same two tRNA genes as does *E. coli*—$tRNA^{Ile}$ and $tRNA^{Ala}$.

Chloroplasts, like bacteria, have only a single type of RNA polymerase complex which contains the same number of subunits as the *E. coli* enzyme, although it is not clear that they are homologous. The drug sensitivity of the chloroplast RNA polymerase is the same as that of bacteria; it is sensitive to rifampicin but resistant to α-amanitin.

Sequence homologies between chloroplast and bacterial rRNA molecules have provided some of the most convincing evidence for the prokaryotic origin of chloroplasts. There appears to be no significant homology between the chloroplast and cytoplasmic rRNAs of an individual algal species. However, chloroplast rRNAs do show significant sequence homology with those of all bacteria studied, with the exception of the Archaebacteria. These homologies are particularly strong with the rRNAs of the blue-green bacteria. The rRNA of the red alga, *Porphyridium*, shows very strong homology with that of the blue-greens and is more closely related to those bacteria than it is to a green alga such as *Euglena*. As described below, there is evidence that the chloroplasts of green algae and of higher plants may have had a number of different origins.

The process of translation in chloroplasts shows many similarities with that in bacteria; protein synthesis is initiated by N-formyl methionine in both and chloroplast mRNA molecules do not appear to be polyadenylated. There is considerable functional homology between plastid and bacterial ribosomes: translational factors may be interchanged between the two systems and hybrid ribosomes containing a chloroplast and a bacterial subunit are functional in *in vitro* translation systems.

The amino acid sequences of a number of important chloroplast proteins, particularly those which are necessary pigments in photosynthesis, have been compared with their bacterial counterparts to determine the degree of homology. Such analyses indicate that the chloroplast ferrodoxins are more closely related to those of the blue-greens than to the ferrodoxins of other bacteria. A complication in this analysis is that the protein component of these pigments is encoded by nuclear rather than chloroplast DNA and so the additional assumption that the relevant gene has been recruited into the nucleus of the host from the proposed endosymbiont must be made. The ferrodoxin sequence data also supports the notion that there was an early divergence of the red algae chloroplasts from those of the green algae and higher plants, a theory supported by the analysis of the amino acid sequences of cytochromes f. Another set of accessory pigments, the phycobiliproteins, are unique to red algae and blue-green bacteria and these show considerable amino acid sequence homology.

Ultrastructural studies of chloroplasts and the membranes which surround them have revealed a further complication to the story of chloroplast evolution. The chloroplasts of all red, and most green, algae are enveloped by two membranes. The inner membrane, according to the xenogenous theory of chloroplast evolution, would have been derived from the cell membrane of the prokaryotic endosymbiont while the outer membrane came from the vacuole of the primitive eukaryotic host. However, the chloroplasts of euglenoids, brown algae and certain green algae have three or four surrounding membranes. It is thought that the chloroplasts of these organisms originated from a eukaryotic, rather than a prokaryotic, endosymbiont and that the extra membranes derive from the eukaryote's endoplasmic reticulum and plasmalemma. These relationships are described in Table 12.

Additional ultrastructural evidence in support of the idea of eukaryotic endosymbionts developing into chloroplasts has come from the discovery of a structure called the nucleomorph in the space between the inner and outer membrane of the chloroplast. The nucleomorph is surrounded by a double membrane penetrated by pores and it is thought to represent the much reduced nucleus of the eukaryotic endosymbiont.

Summary—chloroplast evolution

There is considerable evidence that chloroplasts evolved from free-living bacteria (probably the blue-greens) which established an endosymbiotic relationship with eukaryotic cells. Homology between the genomes of the blue-green bacteria and chloroplasts has been demonstrated and this is particularly striking with the red algae. There are data supporting the multiple origin of chloroplasts and ultrastructural studies indicate that, for some organisms, symbioses between two eukaryotes, rather than between a eukaryote and a prokaryote, may have been involved.

Mitochondrial evolution Our extensive knowledge of the molecular biology of mitochondria demonstrates that they are very diverse, differing not only from each other but also from the rest of the living world. Supporters of the xenogenous origin of mitochondria would contend that this diversity is indicative of their multiple origin. Protagonists of the autogenous route, on the other hand, would

Table 12 The endosymbiotic origin of algal chloroplasts

Number of membranes surrounding chloroplast	Origin of those membranes	Prokaryotic	Eukaryotic
2	Bacterial cell membrane and host vacuolar membrane	Red algae Green algae	
3	Bacterial cell membrane and primary host's vacuolar membrane and primary host's plasmalemma		Euglenoids Dinoflagellates
4	Bacterial cell membrane and primary host's vacuolar membrane and endoplasmic reticulum and plasmalemma		Diatoms Brown algae

claim that it is the result of the elevated rates of mutation found in mitochondrial genomes. The former claim that mitochondria are derived from free-living aerobic bacteria such as *Paracoccus* or *Rhodopseudomonas*, while the latter trace a very ancient line of evolution back to the Archaebacteria. Both arguments have their merits and the data are so complex that it is not possible to make a firm decision either way.

Mitochondrial DNA, like that of the chloroplast, is free of histones. However, it can be either linear or circular in conformation and a tremendous range of both size and base composition is found (see Chapter 2). Many mitochondrial genes, both those encoding proteins and those which specify RNA molecules, contain introns, some of which may be peculiar to mitochondria in their nature and the mechanism by which they are excised. This is probably true of the introns which encode their own splicing enzymes. However, in other cases there appears to be considerable homology between mitochondrial and nuclear introns. An example of this is the homology between certain introns of *Aspergillus* mitDNA and the intron within the nuclear rRNA genes of *Tetrahymena*. The possession of introns, particularly in protein-encoding genes, is regarded as a distinctly eukaryotic feature and this argues against a prokaryotic symbiont giving rise to mitochondria. However, some workers consider that RNA splicing is a very ancient function in evolutionary terms and the recent discovery of introns in some archaebacterial genes confirms this. There is other evidence for a close relationship between the genomes of mitochondria and the archaebacteria.

The organization of rRNA genes in some mitDNA is unlike that found in any organism. In yeast mitochondria, for instance, the 21S and 15S rRNA genes are so far apart that they must be transcribed separately. Another unique feature is the absence of 5S rRNA from all mitochondria except those of higher plants. A search for sequence homology between mitochondrial and bacterial rRNAs has revealed only limited similarities in certain species. This could indicate the separate evolution of eubacterial and mitochondrial genomes or simply be the result of a high mutation rate within mitochondria.

Analysis of the sequences of mitochondrial tRNAs has revealed a number of unique features. These involve all parts of the molecule and include such extreme variations as the complete absence of the 'DHU' loop in some mammalian mitochondrial tRNAs. Sequence homologies between mitochondrial tRNAs and those of the eubacteria or eukaryotic cytoplasm are not strong. However, some fungal mitochondrial tRNAs show a striking resemblance to their counterparts in certain archaebacterial species such as *Halobacterium*, another indication of a very ancient origin for the mitochondrial genome.

Mitochondrial mRNAs display a mixture of prokaryotic and eukaryotic features. They are monocistronic like their eukaryotic counterparts, that is each messenger molecule encodes only a single polypeptide, and mRNAs are also polyadenylated although the poly(A) tracts are very short. The methylated 'cap' structure which is found at the 5′ end of most eukaryotic mRNAs is absent from those of mitochondria.

Mitochondrial ribosomes exhibit a tremendous range of structure and have many features which are unique. They do not contain the 5.8S rRNA which is typical of eukaryotic cytosolic ribosomes, and only the ribosomes of higher plant mitochondria possess the 5S rRNA which is found in the ribosomes of all organisms. Mitochondrial protein synthesis resembles that of bacteria in its use of N-formylmethionine as the initiating amino acid. However, mitochondrial initiator tRNAs have a number of structural features which distinguish them from their bacterial counterparts.

The most striking feature of the mitochondrial translation system is that it uses a different genetic code to the rest of the living world. These differences are summarized in Table 13. Mitochondria appear to have a very limited collection of tRNAs, less than 32 in most organisms, although there is some evidence that these may be supplemented by the use of cytosolic tRNAs. This small number of tRNAs can be employed by mitochondria since they exhibit a distinct codon bias with a universal discrimination against the use of G in the third position. A good example of this bias is the use of UGA (normally a stop codon) to encode tryptophan, which in the rest of the living world is encoded by UGG.

Crick has argued that the genetic code is a 'frozen accident' and that it has remained frozen since subsequent changes would be lethal. This, if true, would indicate that the mitochondrial genome has a very ancient origin with a primitive genetic code (an early 'frozen accident') which uses a limited range of codons. Whatever its origins, the unique nature of the mitochondrial genetic code gives an excellent reason for the continued existence of the mitochondrial DNA. The recruitment of all mitochondrial genes into the nucleus is impossible since those which frequently use unique codon assignments would not be properly expressed by the cytosolic translation system.

Our most detailed knowledge of mitochondrially encoded proteins has come from the study of yeast; these are listed in Table 14. Most theories of mitochondrial evolution invoke the transfer of genes between the nucleus and mitochondrion though they differ on the direction of transfer. It is worthwhile, therefore, to compare which mitochondrial proteins are encoded in the nucleus, and which in the mitochondrion, in different organisms to discover if any evolutionary pattern emerges. The DCCD-binding proteolipid (subunit 9) of the proton-translocating ATPase complex is an interesting case. In yeast, this protein is specified by mitochondrial DNA and synthesized by mitochondrial ribosomes. In the filamentous fungi and in mammals it is encoded by a nuclear gene and synthesized in the cytosol. It has recently been discovered that the moulds *Neurospora crassa* and *Aspergillus nidulans* have a second copy of the gene for subunit 9 in their mitochondrial genomes but that these are not expressed in vegetative cells. *Neurospora* and *Aspergillus* seem to have been 'caught in the act' of gene transfer. The crucial question is unanswered: is the gene being transferred from mitochondrion to nucleus or *vice versa*?

The transfer of genes between organelles represents an additional complication to the evolutionary history of the latter. Two pieces of data have shown that this is

Table 13 Differences between the mitochondrial genetic code and the 'normal' code

Codon	Normally encodes	Encodes in mitochondria from Mammals	*Neurospora*	Yeast
UGA	STOP	TRP	TRP	TRP
CUA	LEU	LEU	LEU	THR
AUA	ILEU	MET	ILEU	probably MET
AGA	ARG	STOP	ARG	ARG
AGG	ARG	STOP	ARG	ARG

Table 14 Proteins encoded by the yeast mitochondrial genome

Three largest subunits of cytochrome oxidase
apocytochrome b
subunit 6 of ATPase complex
subunit 9 (DCCD-binding lipoprotein) of the ATPase complex
Var-1 protein of mitochondrial ribosomes

more than a theoretical possibility. In maize there is a DNA sequence, including regions encoding the 16S rRNA and two tRNAs, which is identical in the chloroplast and mitochondrial genomes. In *Chlamydomonas* the existence of functional interaction between the two genomes is possible, as shown by the indication that mutations which occur in the chloroplast DNA can render both chloroplast and mitochondrial ribosomes resistant to streptomycin. It is evident that the three principal genomes of the eukaryotic cell (nuclear, mitochondrial and chloroplast) are not as isolated from one another as cell structure would suggest but there can be, and has been, extensive exchange of genetic information between them.

Summary—mitochondrial evolution

The great diversity of mitochondria means that it is difficult to propose a unified theory for their evolution. Mitochondria have features which are prokaryotic, eukaryotic and unique. The use of a different genetic code provides a good reason for the continued existence of mitochondrial DNA and hints at an ancient origin. Homologies between mitochondrial and archaebacterial genes are suggestive of a common ancestor. A complete investigation of the relationship between these two groups must await a more detailed understanding of archaebacterial molecular biology.

References

Annals of the New York Academy of Sciences (1981). Vol. 361.

DOOLITTLE, W. F. (1980). Revolutionary concepts in evolutionary cell biology. *Trends in Biochem. Sci.* 5: 147–50.

ELLIS, J. (1982). Promiscuous DNA-chloroplast genes inside plant mitochondria. *Nature* 299: 678–9.

ELLIS, J. (1983). Mobile genes of chloroplasts and the promiscuity of DNA. *Nature* 304: 308–9.

GRAY, M. W. and DOOLITTLE, W. F. (1982). Has the endosymbiont hypothesis been proven? *Microbiol. Revs* 46: 1–42.

GRIVELL, L. A. (1983). Mitochondrial DNA. *Sci. Amer*. 248: 60–73.

WALLACE, D. E. (1982). Structure and evolution of organelle genomes. *Microbiol. Revs* 43: 208–40.

Index

Index